"Compelling and provocative. . . . Should be required reading for all who care about the future of this country and the planet as a whole."

—*The Nation*

"This book is journalism of a high order, unraveling the spin as it leads the reader effortlessly through political and military labyrinths."

—*The Sunday Times*

"[A] very readable account . . . a superb introduction to these issues."

—*The Christian Science Monitor*

"Yeoman does a fine job integrating a wide array of information, from the mechanics of oil production to how the politics of oil helped shape the modern Middle East. . . . Surprisingly nuanced."

—*Newsday*

"An informed primer that weaves together commentary, anecdote, and fact." —*Wired*

Oil

Oil

A Concise Guide to the Most Important Product on Earth

MATTHEW YEOMANS

THE NEW PRESS

NEW YORK
LONDON

Published in the United States by The New Press, New York, 2004
Distributed by W. W. Norton & Company, Inc., New York

LIBRARY OF CONGRESS CATALOGING-IN-PUBLICATION DATA

Yeomans, Matthew.
Oil: a concise guide to the most important product on earth / Matthew Yeomans.
p.   cm.
Includes index.
ISBN 1-56584-885-3 (hc.)
ISBN 1-59558-028-X (pbk.)
1. Petroleum industry and trade. I. Title.
HD9560.5.Y46 2004
338.2'7282—dc22   2004040151

The New Press was established in 1990 as a not-for-profit alternative to the large,
commercial publishing houses currently dominating the book publishing industry.
The New Press operates in the public interest rather than for private gain, and is
committed to publishing, in innovative ways, works of educational, cultural, and
community value that are often deemed insufficiently profitable.

www.thenewpress.com

Illustrations by Scott Lizama
Composition by NK Graphics

Printed in the United States of America

2   4   6   8   10   9   7   5   3   1

For Jowa and Dylan

Energy . . .
Sometimes I think I'm runnin' out of energy
Seems like we use an awful lot for
Heatin' and lightin' and drivin'
Readin' and writin' and jivin'
Energy . . . You'd think we'd be savin' it up.

"The Energy Blues"
—*Schoolhouse Rock, 1973*

Contents

Preface

From the moment we wake up in the morning to the moment we go to sleep, oil controls our lives. Its influence reaches far into politics, international affairs, global economies, human rights, and the environmental health of our planet.

The most obvious way that oil dominates us, of course, is transportation. Oil powers 97 percent of America's transportation needs and over half the oil we consume daily goes to keeping our cars and trucks on the road. That's one barrel out of every seven used in the world. Not surprisingly, the United States has more automobiles than any other country; in fact, it has more cars and trucks than it has people.

But oil is far more important to modern society than simply as fuel for our automobiles and airplanes. Oil provides heat in the winter for millions of American homes, and it accounts for 40 percent of our total energy needs. Without oil there would be no plastics, nor many of the chemical-based medicines we take for granted. Perhaps most important, America would go hungry without oil: commercial agriculture would grind to a halt without oil to run farm and food processing machinery or to make fertilizers, herbicides, and pesticides.

To better understand oil's impact on our lives, I devised a little experiment. I would spend a day without oil. How hard could that be? After all, I live in Brooklyn so I'd already won half the battle; I'd leave the car on the street and hope I didn't pick up a parking ticket.

I began in the bathroom. I'd have to carry off the rough-and-ready look this morning as petroleum products play a role in my shampoo, shaving cream, and deodorant. There was also going to be a lot of water to clear up—my plastic shower curtain is also an oil product.

Brushing my teeth became a far less appealing experience without the benefit of toothpaste, whose ingredients include petrochemical-enhanced artificial coloring and mineral oils. (But at least I still had my own teeth. If I'd worn petroleum-based dentures, I'd be gumming my way through this particular day.)

As it was, I was going to have to make do with only limited vision as both my contact lenses and plastic-lens eyeglasses came from petrochemicals. And I'd have to skip putting on lip balm; that's petroleum oil. Worse still, I'd have to dress my six-month-old son in cloth diapers instead of the normal disposable ones. (What a day to have made the switch to solid food!)

Next came the problem of what to wear. Typically, I live in sneakers but not today—I had to search out an old pair of nonrubber-soled leather shoes. It was raining outside but I had to forgo any waterproof outerwear. Goretex, it turns out, is yet another genius invention of the petrochemical industry.

I left my house and immediately encountered another problem. All New York streets are paved with asphalt, the sticky by-product that remains after refining crude oil to extract its more lucrative properties, like gasoline and heating oil. Lacking powers of levitation, and with not an inch of grass in sight, I had to admit defeat on this point. I traipsed slightly forlorn to my neighborhood café for breakfast. Eggs and coffee came courtesy of a nonstick pan and a heat-resistant glass pot—products of the petrochemical industry. Defeated again. At least I could pay in cash. All credit and debit cards are oil products.

On my return home, I realized this wasn't going to be the most productive day of my working life because I couldn't use the computer or telephone, both of which are housed in plastic. Neither could I kick back and listen to music or watch a movie—CDs and DVDs also contain oil. Perhaps then I could just go and play a round of golf? Stuck again: golf balls contain polybutadiene, another petrochemical.

The list of off-limit items continued. Bandages, blenders, garbage bags, glue, pacemakers, and pantyhose (the latter two not being items I needed on this particular day) all got their start as oil. This whole day-without-oil thing was beginning to give me a headache. Perhaps I should just take a few aspirin and forget about the whole thing. You guessed it: Aspirin is another proud legacy of oil.

★   ★   ★

No nation uses more oil than the United States. The U.S. uses a full quarter of all the oil produced in the world. And even though it produces nearly eight million barrels of oil a day, it must import a further twelve million barrels just to meet its daily needs. Yet despite this amazing discrepancy, we still take oil for granted.

How could it be that a people whose way of life is so dependent on oil could be so naive about how we use it? The more I thought about it, the more I began to realize that America's oil addiction runs deeper than a love for big cars and the open road.

Oil, I believe, has etched an indelible yet invisible mark on the modern American psyche. After all, America is the birthplace of the oil industry. Cheap, plentiful supplies of U.S. oil propelled the United States to victory in two world wars. Oil fed the post–World War II consumer boom that established a consumption-crazy middle class. And the oil industry helped spawn another industrial giant, the automobile industry. At every step of the way, oil has provided a blueprint for modern American life that the rest of the world eagerly copied. Americans, for their part, have embraced the spirit of self-confidence and invincibility that grew from knowing they controlled their own oil destiny. And that spirit continues today, even though America has long depended on others for oil.

One weekend last summer, I took a drive out from New York to the rolling hills of western Pennsylvania. I wanted to visit the birthplace of the oil industry. I wanted to see where the roots of modern life began. And I wanted to capture a bit of the old oil spirit.

Six hours' drive west of New York, I pulled off Interstate 80 and headed north following the Allegheny River as it winds its way through some of Pennsylvania's most rugged and beautiful countryside. Climbing up through the hills, I came to the small town of Oil City, its impressive late-nineteenth-century stone buildings a proud reminder of the time when this rural patch of Pennsylvania was the very heart of the world oil industry.

Oil City was once the most important boomtown of what became known as Pennsylvania's oil regions, a center of oil exploration that sent crude oil north by rail to refineries in Cleveland, Ohio. In the early 1860s, 2,000 oil wells dotted the land around the town and the entire area resembled an oil derrick pincushion. For a short time, Oil City was as famous as Houston.

In 1857, a small group of New Haven, Connecticut investors founded the Pennsylvania Rock Oil Company to tap into a flammable substance that seeped from the earth around the hills in this part of Pennsylvania. "Rock oil" was being used by the locals to clear headaches, toothaches, and even upset stomachs.

The investors were interested for another reason. Global supplies of the whale oil that had long been burned in lamps were dwindling as sperm whales had been hunted to the point of extinction. Meanwhile, in Europe, a new process was being employed to dig this rock oil out of the ground by hand so that the stuff could be refined to make a new illuminant, kerosene.

The Pennsylvania Rock Oil Company had a better idea for extracting oil—it would drill for it using the same techniques already used at salt wells around the country. The man the investors chose to lead this venture was Edwin Drake. The thirty-eight-year-old jack of all trades was between jobs at the time, but he happened to be living in the same hotel as one of the main investors. The company decided on him partly because he portrayed himself as a can-do type of guy and partly because, as a railroad conductor, he could travel to Pennsylvania for free.

To bolster Drake's credibility in Pennsylvania, the company sent letters ahead of his arrival addressed to "colonel" E.L. Drake. The title stuck and, in 1858, the Colonel began drilling a few miles south of the little hamlet of Titusville, some fifteen miles north of Oil City, on a farm that contained a seeping oil spring. For over six months he had dug deep but dug dry. He was still searching for oil and had exhausted all the funds of the investors when, out of desperation, they mailed him instructions to abandon the operation. But the letter didn't reach Drake until after he had tried one final well. This time, the drill bit sank sixty-nine feet into the ground—then slid six inches more. Drake had hit oil and the industry that would change the world was born.

Soon thousands of prospectors—many of them returning from the Civil War—were flocking to the area around Titusville in search of their fortune. No place better epitomized the oil region's frenzy of speculation than the town of Pithole, a few miles south of Drake's first discovery in Titusville. At its peak in 1865, Pithole had 10,000 inhabitants, fifty hotels, two telegraph offices, and a post office that was said to be the third busiest in the United States. It was also a haven of prostitution packed with fifty "free and easy" brothels.

But Pennsylvania's dominance was short-lived. Within a year Pithole was gone. The oil fields had been pumped too quickly by the prospectors, and once abandoned, Pithole's wooden buildings burned to the ground in a fire. Soon, bigger and better oil fields were discovered in Texas, California, and Oklahoma, then farther afield in Russia, Mexico, Venezuela and eventually, in the desert lands of the Persian Gulf. In the process, big oil companies—most of them American—would ride the booms and the busts and grow so rich that they often became more powerful than the countries where they drilled for oil.

While the rest of the state's oil industry suffered, Oil City received a second lease on life in the 1920s when a company named Pennzoil relaunched the town's oil business. Oil City became a company town—its livelihood totally dependent on Pennzoil's business. Then, in 1995, the company pulled the plug on Oil City and moved to Texas where production costs were cheaper. Today, as you walk down the once vibrant Seneca Street, you can't help but notice that most of the stores have been replaced by boarded-up buildings. Even the town's central landmark—the Oil City Savings Bank—now opens only for a weekend flea market. "This town used to be full of nice shops and boutiques," said the owner of the Yellow Dog Lantern restaurant when I stopped by for dinner. "Now there is nothing left."

Oil City discovered the perils of building a society that is totally dependent on oil. It is a lesson that America as a whole has yet to learn. Yet the problem wasn't that the oil region ran out of oil. There are still some 4,000 active small oil wells across the state. No, Pennsylvania's problem, one that Oil City was able to put off for a while thanks to Pennzoil's business, was that it could no longer compete against newer, cheaper sources of oil from other parts of America and the world. Oil fields rarely run dry, but as they get older—or grow more mature, as the oil industry calls it—the oil becomes more expensive to produce.

It wasn't until the 1970s that the U.S. first realized it had an Achilles' heel. The 1973 Arab oil embargo, coupled with a series of price hikes by the Organization of Petroleum Exporting Countries (OPEC), shocked the U.S. into realizing that its oil dependency had become a seriousness weakness.

Yet since that time, oil production in America has fallen rapidly. U.S. oil fields in the lower forty-eight states are 25 percent less productive today than they were in 1985. And since 1990, proven oil reserves have dropped 20 percent. Unable to meet its own needs, and wary of being burned once more by OPEC, the U.S. has made a concerted effort to spread its oil imports across a large number of producing countries. Winston Churchill once said that oil "security exists in variety and variety alone" and this has been the mantra of U.S. governments since the end of the 1970s.

The U.S. presently gets just 25 percent of its imported oil from the Middle East, and it has cultivated important relationships with Canada, Mexico, and Russia as well as other producing nations in Latin America and Africa. Europe and Japan, with little oil of their own to draw upon, have also sought to diversify their oil imports, though they still rely more on Middle East oil than the U.S. does.

But despite this emphasis on diversity, everyone associated with the oil business knows that the future of oil will be decided by one region, the Middle East. For all the new technology and all the financial resources that oil companies can call upon to find and drill oil around the world, the Middle East has not just the most abundant and deepest reserves of oil in the world but also the oil that is cheapest to produce. As the lessons of Pennsylvania's oil regions show, it is only a matter of time before the rest of the world's oil becomes uncompetitive compared to that of the Middle East. When that happens, Saudi Arabia, Iran, and Iraq will be able to set their price.

For as long as we are dependent on oil, we will face a triple threat.

The first part of the threat is economic—as the balance of power shifts once more to the Middle East, America's oil dependency is likely to leave the nation once again at the mercy of OPEC. At that point, oil prices are likely to start rising and that could threaten the jobs of millions of Americans and ultimately undermine the economic well-being of the country.

The second is geopolitical—the United States must protect its oil interests all over the world. That forces the U.S. to get directly involved in those regions where oil is abundant, regions where the very power and allure of oil wealth breeds instability. The end result is that the U.S. supports a host of corrupt autocratic regimes solely because they have oil. At the very least, this fosters anti-American sentiment. At its worst, America's oil addiction has caused U.S. governments to plot the overthrow of other countries' governments. Increasingly, the U.S. is shaping a military policy based on protecting strategic oil resources around the world.

Finally, oil dependence is sowing the seeds of a global environmental tragedy. Even if the U.S. could produce enough oil to meet its own needs, or could solve all the problems of the Middle East, the rapidly growing demand for oil throughout the world will result in the large-scale release of greenhouse gases like carbon dioxide into the atmosphere. Global warming—an assault on the earth's atmosphere the scale of which no one in the world's scientific community can yet fully comprehend—could ultimately destroy the very planet that sustains us.

So is all lost? Should we just give up now, buy Hummers and drive off into the brightly glowing and rapidly warming sunset? Hardly. Our dependence on oil grew over just 150 years. That's a mere speck in time in the history of our world and it is also the best indicator that our addiction may also be fleeting. Oil replaced our reliance on more polluting fossil fuels like wood and, later, coal. Now, having appreciated both the achievements of oil and its risks, we can move to a new energy strategy that would both sustain us and protect us in decades to come. With global demand for oil likely to double by 2030, the time has come to develop a new fuel strategy that reduces the threat of global warming and eases the power struggles that threaten our security.

This strategy should be realized in three stages. First, we in America should reduce the amount of oil we consume by implementing new fuel economy standards for our automobiles. Next we should adopt a new generation of gasoline-and-electric-hybrid vehicles that can provide a bridge between the internal combustion engine and new, nonpolluting forms of transportation. Finally, we should put our resources behind an aggressive plan to make hydrogen the fuel that powers our automobiles. Most important of all, this hydrogen needs to be produced using nonpolluting, renewable energy forms like solar, wind, or hydroelectric power.

I'd been writing stories about oil for nearly a decade before I fully began to appreciate its power. At first, I viewed oil as just another big industry, albeit one ruled by some of the largest multinational companies on earth. But over time I started to realize that oil was a lot more important than just a business.

The main reason that it took me so long to appreciate the power of oil was because it flowed so deeply through my life. This was the industry that had built those cool supertankers that mesmerized me as a child growing up on the coast of Wales, and the same industry whose gas stations gave away coin collections commemorating all my favorite English soccer teams. For a while, my family even made special trips to one Esso gas station to collect free drinking glasses (I think my mother is still using them). As a teenager, I was taught how

North Sea oil was the lifeblood for a British economy that was sinking otherwise into a pit of moribund industry. And on my first visit at age fifteen to the U.S. I fell in love with American cars and car culture. Life in America seemed so fast, so supercharged, compared to back home, and it left an indelible mark on my teenage mind—so much so that I would later settle in New York.

At the same time, this was the same industry that was responsible for an oil spill that wrecked the coast near my home. It was the industry that flooded Alaska's Prince William Sound with millions of barrels of oil from the Exxon Valdez within months of my arrival in the U.S. Later, I would see the full extent of oil pollution and destruction when I traveled the fringes of the Amazon rain forest and met whole communities whose lives had been ruined by oil.

These contradictions had filtered through my brain for years as I followed the oil industry from afar, but it wasn't until the months immediately following the September 11, 2001, terrorist attacks on New York and Washington, D.C. that I started to consider the larger role of oil on our global society.

I started to notice how so many of the stories I read each morning in the newspaper were connected to oil. There were the obvious ties between Osama bin Laden's anti-U.S. jihad and America's military presence in Saudi Arabia—a hangover of the Gulf War and U.S. commitment to protect Middle East oil supplies. And there was the increasing military deployment of U.S. forces in the countries that surround the oil-rich Caspian Sea. Then there were the environmental and energy policies of the Bush administration, which sought to pump new life into America's domestic oil industry while rejecting many of the environmental controls that the rest of the world seemed to agree could combat global warming.

And then there was Iraq, and the U.S. government's bellicose drive to overthrow Saddam Hussein. Opponents of the Bush administration inside the U.S. and all over the world were convinced that this was a war over oil. At the same time, both the U.S. and the UK government of Tony Blair were adamant that oil played no role in the push to putsch Saddam.

The Bush administration's preemptive war seemed both preposterous and crass even without the benefit of hindsight that we now enjoy. But I couldn't buy the no-blood-for-oil argument. The U.S. had spent the last twenty years cultivating relationships with other producing nations and it didn't have to invade Iraq just to satisfy its own oil needs. However, I was convinced that oil played some crucial role in the geopolitical maelstrom surrounding Iraq that neither the pro- nor anti-war crowd was addressing.

That was when I decided to undertake this journey through the world of oil. The purpose of the journey was simple—to understand and explain in straightforward terms the ways that oil has come to control our lives.

Acknowledgments

Oil, and the hold it maintains over the world, is an enormous topic. My goal in this book was to condense and succinctly convey available research on the subject in an informative and easily digestible manner.

A lot of people have helped me in this goal. My thanks go to Andy Hsiao, my editor at The New Press, who helped me shape the book before I began writing and then expertly trimmed my rambling, often jargon-ridden prose. Also at The New Press, my thanks to André Schiffrin and Colin Robinson for giving this project the go-ahead, Steve Theodore and Maury Botton for making sure it stayed on the rails, and Izzy Grinspan for her fact-checking help. Scott Lizama did a great job translating my rough descriptions into clear, precise maps and illustrations.

I was lucky to call on the expert eyes of a number of friends and colleagues during the research and writing of this book. My agent, Kay MacCauley, is always full of encouragement for my ideas. Thomas Goetz offered advice and a close read from first to final draft. Mark Wise pointed me in the right direction when my copy was rough. Scott Malcomson fine-tuned my prose when it most needed it, while Steve Kretzmann and Shannon Wright offered feedback on various chapters. Rob Boynton did double duty: he read parts of my manuscript and he let me use his office just when my newborn son made working from home a little distracting.

I'd like to thank the Journalism Department at NYU, where, as a visiting scholar, I was able to call upon the wealth of its research tools. Thanks also to Grand Valley State University in Michigan, whose campus center became my de facto office during the summer of 2003 when I escaped from New York to write the book.

Special thanks to my in-laws, Rob and Lavonne Coffey, who opened their home to me during those summer months. Thanks also to Rob for initiating me into Michigan car culture and explaining the meaning of a Hemi. Finally, my love and thanks go to my wife, Jowa, who provided the daily support and encouragement I needed to complete this book, and to my son, Dylan, who understands nothing about oil but makes me smile every single day.

Any factual mistakes that arise from crunching all this material are entirely of my own doing. However, some information that was accurate at the time of writing will have changed by the time you come to read it. Such is the fluid nature of the politics and economics of oil.

Matthew Yeomans
Brooklyn, New York
February 2004

1

Pursuit of Power:
A Short History of Oil

The prospectors who descended on Pennsylvania's western hills in the months following Drake's 1859 discovery of "rock oil" had no idea of oil's true potential. They didn't know that oil would dictate the fortunes of millions of people for the next 150 years. They didn't know that the pursuit of oil would push great nations into war and they didn't know that their initial wildcatting would propel the United States to become the most important nation on earth. All that mattered at the time was that there was a fortune to be made. Kerosene lamps were already being celebrated as "the new light" in cities up the eastern coast and Pennsylvania oil promised a new and abundant source.

Thousands of people flocked to stake a claim in the new oil regions. In January 1861, barely a year after Drake's initial discovery, the price of a barrel of oil hit $10 (Drake had collected the first oil in whiskey barrels and the forty-two–gallon measure has remained the industry standard ever since). By the end of the year, however, as more and more wells were struck and prospectors frantically offloaded their oil, the new pioneers came to realize just how fickle their fortune could be. Prices dropped to 10 cents a barrel and the oil industry confronted what would be the first of many gluts in its history.

Yet out of the oil regions' chaos came order and control in the figure of John D. Rockefeller, owner of a small refinery in Cleveland, Ohio. Rockefeller observed how the hundreds of Pennsylvania drillers were bidding against each other and so were driving down the cost of oil. Rockefeller realized the essence of making money in this new volatile industry lay not in producing oil, but in controlling the transport and sale of oil's refined products.

Though only twenty-six years old, Rockefeller bought out his refinery partners and started building up a strong cash position. This liquidity gave him the leverage he needed to buy out competing refineries when the next production glut sent oil prices plummeting and cut revenues for both producers and refiners.

Rockefeller soon began undercutting his rivals by persuading the railroads that carried the oil out of the oil regions to give him secret rebates on his shipments based on the volume of his business. By 1870, Rockefeller was able to establish a joint-stock company named the Standard Oil Company. After just seven years in the business, this company controlled one-tenth of the U.S. oil industry.

Standard Oil's business methods were ruthless. The company tried to squeeze many competitors out of business by deliberately cutting its prices in one market, knowing all the time that with its superior cash reserves it could hold out until competitors folded. When Rockefeller bought these companies, he would keep the transactions secret so that his other competitors would not know which companies were owned by Standard and working against them.

By 1883, Standard was laying the foundations for what we now know as the vertically integrated company and the modern multinational. Rockefeller owned refineries and the new pipelines that were transforming the transportation of oil. He had bought up oil fields. And he was opening up new markets for his products on the East Coast and Europe.

In 1906, following years of investigations and a series of damning newspaper and magazine articles on Standard's predatory practices, the U.S. government brought a massive lawsuit against the company under the statute of the 1890 Sherman Antitrust Act. By now, Standard had recorded over a billion dollars' profit since its inception twenty-five years before. The case dragged on for five years and Standard argued it all the way up to the Supreme Court. But in May 1911, Chief Justice Edward White declared Standard Oil a monopoly and ordered that it divest itself of all its subsidiaries.

By the end of the year, Standard had been split into thirty-four companies. The breakup destroyed Standard's singular stranglehold on the oil industry but it didn't destroy the strength of the Standard brand. On the contrary, Standard's model was so strong and the U.S. oil industry so vibrant that a number of the new mini-me Standard clones soon emulated Standard's success. Three in particular, Standard Oil of New

SONS OF STANDARD OIL: THE SUPERMAJOR FAMILY TREE

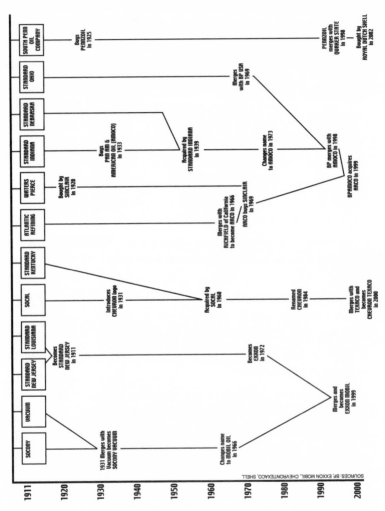

SOURCES: BP, EXXON MOBIL, CHEVRONTEXACO, SHELL

Jersey (later to be known as Exxon), Standard Oil of New York (Socony and later Mobil), and Standard Oil of California (Socal, later Chevron) quickly became global powers in their own right.

Enter Europe

News of American oil quickly spread to Europe and soon Standard Oil was contracting more and more ships to meet this international surge in demand. In the 1870s and 1880s, kerosene exports—most to Europe—accounted for half of all U.S. oil production and provided the fourth-largest U.S. export in value. Ninety percent of that kerosene came courtesy of Standard Oil.

Europe also quickly realized that oil was not purely an American gift. For centuries, travelers had talked of flaming pillars that burned in Baku on the Caspian Sea and by 1871, the Russian town of Baku was dotted with oil derricks. Russian oil attracted the Rothschilds, a prominent French banking family, into the European kerosene industry, but they found their path blocked by Standard Oil. Undetered, the Rothschilds decided to sell oil in Asia, where Standard's grip wasn't as complete.

The man the Rothschilds found to successfully challenge Standard Oil in Asia was Marcus Samuel, a Jewish merchant from London who, thanks to his father, an import-exporter and trader of handmade seashell gifts, had strong shipping connections throughout the Far East. Samuel knew that to succeed against Rockefeller he would have to outlast Standard's practice of flooding the market with cheap oil. To do that, he would have to sell in all Asian markets at the same time, ensuring that Standard couldn't concentrate on just one market. But that meant Samuel would have to get his Russian oil to the Far East before Standard and its large network of international corporate spies found out.

In fashioning a solution, Samuel invented a new way of transporting oil and established one of the industry's most important companies. First, rather than shipping barrels of oil, Samuel commissioned the construction of a new ship—one that carried the kerosene in an enormous tank. Second, he cut the sailing distance of these tankers in half by getting safety clearance from the British government to navigate the new Suez Canal (kerosene shipments had been banned due to the fear of the vessels exploding in the canal), thereby avoiding the long trip around Africa. In 1892, Samuel's first tanker, the *Murex,* named for a type of shell, sailed through the canal. Within a decade, his new company, Shell, controlled 90 percent of the oil passing through the Suez Canal.

The Start of the Addiction

Perhaps it was because of Samuel's understanding of maritime trade, or maybe it was because Shell now had large excess reserves of Borneo fuel oil following its recent merger with Royal Dutch, an East Indies–based rival, but Samuel led the campaign to make the British Royal Navy abandon coal-fueled ships in favor of oil.

Since the late 1890s, Britain and Germany had been caught up in an increasingly high-stakes naval arms race. Central to both sides' military ambitions was control of the world's oceans. The British admiralty was dead set against switching to oil—not least because, while Britain enjoyed great reserves of coal, it had no oil of its own. Nevertheless, after a decade of lobbying, Samuel finally got the ear of a new First Lord of the Admiralty, the young Winston Churchill.

Churchill quickly grasped the advantages of an oil-fueled battle fleet. Between two such well matched imperial powers as Britain and Germany, naval superiority would tip the balance and Churchill saw how Britain could achieve that. Oil fuel allowed faster cruising speeds and faster acceleration than coal furnaces. It took up less room allowing for greater armaments and manpower. It was also cheaper to operate. In April 1912, Churchill took the "fateful plunge," as he described it, and commissioned a series of new battleships all dependent on oil.

In one swoop, Churchill had made the security of Great Britain dependent on foreign oil. Yet the lure of oil—its mobility and the advantage it afforded the British fleet at war—was enough to persuade him. With oil, Churchill wrote, "we should be able to raise the whole power and efficiency of the Navy to a definitely higher level; better ships, better crews, higher economies, more intense forms of war power." As he put it, "mastery itself was the prize of the venture."

Churchill found the oil he needed to run his navy in Persia where a new company, Anglo-Persian Oil, had struck a rich seam of oil. Yet despite the company's potential, it was desperately short of capital. In 1913, Churchill announced that, in the interests of national security, the government would buy 51 percent of the company and Anglo-Iranian would sign a long-term contract to supply fuel oil to the British navy. The agreement stipulated that the company must always remain a British concern and, to protect its investment, the government increased its military presence in Persia. Anglo-Persian would later change its name to Anglo Iranian, then British Petroleum, or BP. And Great Britain had become the first Western power to tie its economic and national security to Middle East oil. Others would soon follow.

The First Oil War

Nowadays we take technology in warfare for granted. Ever since the first Gulf War in 1991, the media has fallen over itself to catalogue new military inventions—be it stealth fighters, smart bombs, or real-time video footage of the battleground that can be monitored from thousands of miles away. But, as Germany and the Allied powers of Great Britain and France squared off in 1914, neither side understood the differences that the internal combustion engine would bring to modern conflict.

World War I dragged on in an increasingly bloody stalemate for four years as each side introduced more deadly, mobile weapons and the carnage grew exponentially. In 1916, the tank was introduced to combat and by 1918, the Allied forces were using over 150,000 cars, trucks, and motorbikes to transport troops and supplies. During the course of the war, the combustion engine also took to the skies. In 1915, the Royal Air Force had only 250 planes to call upon; by war's end, British industry had produced 55,000, France 68,000, and Germany 48,000.

These inventions increased mobility on the battlefield, spreading the conflict over a far greater area than military planners had ever imagined. And it changed the odds of warfare. Even the finest infantry and cavalry were no match for the new fighting machines. Over thirteen million people died and millions more were wounded during the four-year conflict.

It took huge quantities of oil to supply both sides' war effort. Oil production at Anglo-Persian's operations increased tenfold from 1912 to 1918 and in 1917, Great Britain, with an eye to Mesopotamia's oil potential, captured Baghdad from the Turks. Yet Britain and France still found themselves facing huge oil shortages at the height of the war. There was only one place they could turn for help—the United States. By 1917, the U.S. was producing 335 million barrels of oil, 67 percent of total world output, and nearly one-quarter of that was sent to Europe. In total, the U.S. supplied 80 percent of the Allies' wartime petroleum needs. One-quarter of that came from Standard Oil of New Jersey.

Germany's oil problems were even more severe. Cut off from overseas oil by the Allied naval blockade, it had only one other option—the oil fields of Romania. Yet despite a full-press effort to capture the oil fields, British saboteurs got there first, putting Romanian oil out of action for five months. On November 11, 1918, Germany, faced with an acute oil shortage for the coming winter, surrendered. As Lord Curzon, a member of Britain's War Cabinet, triumphantly announced, "The Allied cause had floated to victory on a wave of oil."

The Rise of Big Oil

The lessons of World War I were all too obvious—the great powers needed oil to survive. And while new finds were being made in the U.S. and Mexico, one region of the world was attracting more interest than any other—the Middle East. Prewar the U.S. had been content to develop its domestic reserves and leave foreign prospecting to the Europeans. Now, it was adamant that Britain and France should not carve up the Middle East alone.

Britain and France had agreed to divide up responsibility for the Arab lands of the Ottoman Empire, which had been allied with Germany during the war, but whose alliance had now disintegrated. Central to their ambitions was Mesopotamia, which Britain quickly renamed Iraq, and specifically the regions around Baghdad and Mosul that were believed to hold huge oil reserves. Iraq's potential had first been identified by a savvy Armenian oil prospector, Calouste Gulbenkian. In 1914 he put together the Turkish Petroleum Company, a syndicate involving Anglo-Persian, Shell, Deutsche Bank, and himself, holding 5 percent of the concern.

After the war, Britain and France agreed to let Standard of New Jersey (Exxon), Socony (Mobil), and five other U.S. companies take Deutsche Bank's stake in what was now called the Iraq Petroleum Company. With Britain overseeing Iraq, the local puppet rulers rubber stamped a very favorable concession for the foreign companies—exclusive drilling rights in Iraq until the year 2000.

Gulbenkian, wary of being squeezed out by the major oil companies, was insistent on one condition in the contract: no member of the Iraq Petroleum Company would undertake operations anywhere else within the Middle East lands of the former Ottoman Empire without the joint cooperation of the other members. There was just one problem—no one was really sure just how far the Ottoman lands extended. So when the new members of the company finally came together to finalize the deal, Gulbenkian took a red pencil and drew a rough line on a map around what he understood to be the Ottoman lands. Inside the line lay all of Iraq and Saudi Arabia. Gulbenkian's guess had fortuitously pinpointed the greatest oil fields in the world and it put them in the hands of Western oil companies. In establishing a de facto non-compete agreement—between Shell, Anglo-Persian, Standard Oil of New Jersey, and Socony (along with France's national oil company)—Gulbenkian's Red Line agreement, as it became known, created a cartel out of the world's preeminent oil com-

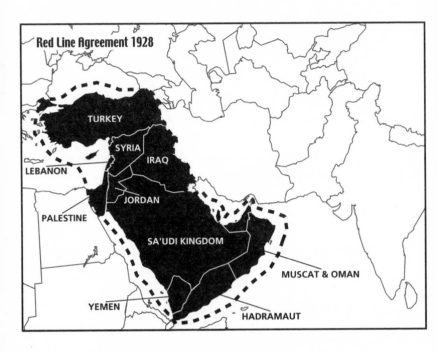

panies. As for Gulbenkian, he got to keep his original stake in the company. He would forever be known as Mr. Five Percent.

The Second Great Oil War

Twenty years after the end of World War I, Europe once again found itself facing war. This time, all the powers knew that oil would make the difference between victory and defeat. The U.S. once again used its plentiful oil fields to supply its own war needs and those of Great Britain. But Germany, just as in World War I, was forced to undertake a risky strategy to capture foreign oil supplies.

Germany believed that its key to victory was a series of short overwhelming mechanized attacks—the blitzkrieg—and at first it was very successful. One reason for the blitzkrieg was that Hitler knew he didn't have enough oil to compete in long, drawn-out battles. To counter this weakness, Germany had two goals in mind—it's Panzer tank divisions would punch through Russia and snatch the Baku oil fields before continuing on to secure the grand prize of Iraq and Iran. But Operation Blau, as this grand oil grab was called, faltered before it could ever reach Baku. The German army had to travel thousands of miles to reach the Caucasus and the speedy Nazi tanks outran their own fuel supply lines. Short of gas, and caught in the heavily defended Caucasus mountains, the armored divisions had to be refueled by camel trains as their own trucks had run out of gas. In its desperation to capture Baku, Germany left its Sixth Army stranded and short of fuel outside Stalingrad.

Surrounded by Soviet forces after a six-month siege, the Germans needed to fight for only thirty miles to escape. But their tanks only had twenty miles of fuel in them. The decision to strand the Sixth Army by refusing to divert the Baku-bound forces was taken by Hitler himself. "Unless we get the Baku oil, the war is lost," he told his commander of the forces in the Caucasus.

Fuel shortages would continue to bedevil both sides as the war went on. General Erwin Rommel, father of the Afrika Korps, the most mobile and effective tank division Germany possessed, was undone in North Africa when the Allies destroyed his refueling lines of supply while the hard-charging American General George Patton and his rampant Third Army were prevented from what could have been an early and decisive invasion of Germany in 1944 by a lack of fuel. "My men can eat their belts," he said, "but my tanks have gotta have gas."

The United States might never have entered the war had it not been for Japan's desperation to capture the oil fields of Indochina. As early as the mid-1930s, Japanese economic planners had come to the conclusion that Japan's own plans for aggressive colonization of Southeast Asia would fail unless it controlled its own oil destiny. That meant invading the oil fields of the Dutch East Indies, but this strategy risked an attack by the U.S., which, already wary of Japan's imperial ambitions, had recently moved the American fleet from California to Pearl Harbor in Hawaii. Faced with Japan's invasion of southern Indochina and its new alliance with Germany and Italy, the U.S. had frozen all Japanese financial assets, cutting off its ability to purchase U.S. oil.

It was at this point that Japan took its own fateful plunge—the preemptive attack on Pearl Harbor in an attempt to destroy U.S. influence in the Pacific. To a degree they were successful. The December 7, 1941, assault decimated many ships in the fleet. But the Japanese also made a big mistake. They failed to hit the four and a half million barrels of oil stored at Pearl Harbor. If they had destroyed America's Pacific oil reserves the whole fleet would have been immobilized.

Over the next four years, Japan would be methodically expelled by Allied forces from the Pacific islands and Southeast Asia. And with each defeat, it saw its access to oil dwindle. By the time the first atomic bomb was dropped on Hiroshima in August 1945, Japan was already a spent force. The U.S. navy was sinking every Japanese oil tanker before it could return home from the Dutch East Indies and the Japanese navy didn't have enough fuel to leave its home base. The shock of the atomic attacks ended the war, but it was a lack of oil that defeated Japan.

The Age of Energy

The war in Europe had only been over for a few months when President Roosevelt traveled to meet a world leader he might not have given the time of day to just a few years before—Saudi King Ibn Saud. There is no official record of what Roosevelt and the Saudi king discussed, but a number of accounts assert the U.S. president promised to guarantee Saudi national security (not least from Great Britain, whose imperial traits Ibn

Saud had a particular dislike for) in return for continued and preferential access to Saudi oil. The potential prize was huge—in 1939, Standard Oil of California along with Texaco had struck Saudi oil. The companies had agreed to a sixty-year concession with Ibn Saud covering 440,000 miles, one-sixth of the continental U.S. Now, geologists at the newly formed Saudi-American Oil Company, Aramco, confirmed there was more oil underneath the Saudi sands than in the whole of the United States.

The U.S. and Saudi Arabia had good reason to be concerned. The world was divided into two camps and Middle East oil was caught in a superpower struggle between the U.S. and the Soviet Union. As Daniel Yergin describes in *The Prize,* his sweeping history of the oil industry, the U.S. believed that "the Middle Eastern oil fields had to be preserved and protected on the Western side of the Iron Curtain to assure the economic survival of the entire Western world."

In 1940, America had accounted for two-thirds of all global oil production. But the immediate postwar years saw an enormous growth in world oil demand and, while the U.S. could still meet most of its own domestic demand, Saudi oil soon became an important part of the Marshall Plan to bail out Western Europe. President Harry S. Truman wrote to Ibn Saud in 1948, "No threat to your Kingdom could occur that would not be a matter of immediate concern to the United States."

Socal and Texaco had won the right to develop Saudi Arabia's oil industry but they quickly realized that the task was so immense that they needed outside help. The companies that

Socal and Texaco reached out to were Standard Oil of New Jersey and Socony.

The entry of Standard Oil of New Jersey and Socony into Saudi Aramco signaled the end of the Red Line agreement, but it also allowed all the U.S. and European majors to put aside their rivalries and protect their mutual interests in the Middle East. United, they could shut out other oil companies from lucrative contracts and also control Middle East oil output. So successful were they in this strategy that they earned themselves the enmity of other independent companies. They also picked up a sardonic nickname to describe their cozy relationship, the Seven Sisters. Through the multi-decade sweetheart deals signed in Iraq, Iran, and Saudi Arabia, the Sisters (Standard Oil of New Jersey, Socony, Shell, BP, Texaco, Socal, and Gulf) would transform themselves into the greatest multinationals in the world. For a time, they would be more powerful than the countries whose oil fields they were drilling.

The U.S. government let the four Aramco members act as de facto U.S. ambassadors to Saudi Arabia, not least because the U.S. government was a strong supporter of the new State of Israel, a position the anti–Zionist Saudi king abhorred. Despite their disagreements over Israel, both the U.S. and Saudi Arabia found common cause in a greater concern: the Soviet Union's own territorial ambitions in the region. And this wedded U.S. foreign policy even more closely to the business needs of the major oil companies. So while back in the States, the companies were fighting tooth and nail against the government over new antitrust and price-fixing charges, abroad they were America's eyes and ears in what was becoming the most influential region of the world. This blurring of business and diplomacy suited both parties: the oil companies had been negotiating deals in the Middle East since the 1920s and understood the workings of the region better than the U.S. State Department. The companies' close ties to the U.S. government also enhanced their standing with Middle East governments. Before long though, the perception that the majors were doing Washington's bidding would come back to haunt them.

Oil Bites Back

A generation had passed since the oil companies brokered their first Middle East oil deals. World oil consumption had exploded during the immediate postwar years and the Middle East was the source of much of that growth. None of this was lost on a new set of leaders in the Middle East who resented the carve-up of their region by the colonial powers and the cut-price concessions the major oil companies had negotiated. In a world dependent on hydrocarbons, the Middle East could and should get a greater share of the profits, the producing nations reasoned. The time had come to challenge the majors.

The catalyst for this challenge came not from some cavalier Lawrence of Arabia figure or even from Soviet meddling. It came from Venezuela, where, in 1948, a populist government had passed a new petroleum law. It ensured Venezuela would now share all "rents"—the market-share profits plus extra fees that took into account various costs of production—in a fifty-fifty split with the major oil companies. The companies, realizing that the U.S. government wasn't going to back them up and fearful that they might lose everything if Venezuela nationalized its oil business, agreed to the new deal.

Word of the Venezuelan deal spread, and by 1949 Saudi Arabia was demanding the same deal. Aramco might have rebuffed the Saudi government had it not been for a new group of independent oilmen, epitomized by Oklahoma millionaire J. Paul Getty, who offered far higher terms for new Saudi concessions. If this American was prepared to pay so much, then obviously the majors were taking the Saudi government for a ride. Soon, Kuwait and Iraq had also cut their own fifty-fifty deals.

Iran also tried to get BP, the sole operator in the Anglo-Iranian Petroleum Company, to agree to such a deal, but its chairman, William Fraser, rejected it outright. This rejection would prove disastrous for BP. In 1951, the new prime minister of Iran, Mohammed Mossadegh, called for the nationalization of Iranian oil and the seizure of BP's oil fields. BP retaliated by organizing a boycott of Iranian oil, effectively depriving Iran of its primary source of income.

Mossadegh's revolt was a direct affront to Western interests in the Middle East. If BP could be thrown out of Iran, what would that mean for the other majors in the region? The U.S. and Britain may have been concerned about the Soviet threat in the region, but they were more worried about their companies losing their sweetheart deals. So, in a move that continues to affect Iranian attitudes toward the West today, the CIA, at Britain's prompting, staged a coup and forced Mossadegh out of office. In his place they put the young Mohammed Reza Pahlevi on the Peacock Throne. The West, it was clear, would let no one interfere with its control of Middle East oil.

Mossadegh's message of resistance was carried on by Egyptian dictator Col. Gamal Abdel Nasser. He was not just a nationalist and anticolonialist; he also sought to unite the Arab world in a campaign for the dissolution of Israel. Nasser was perfectly candid about the role oil played in his revolutionary thinking, calling it "the vital nerve of civilization" and vowing to use oil as a weapon to overcome imperialism.

The only problem for Nasser was that Egypt didn't have any oil. But it did have the Suez Canal, which carried the majority of Middle East oil shipments to Europe even though stewardship of the canal was still controlled by Britain and France. In just a few months in 1955, Nasser successfully scared the hell out of the U.S. and Western Europe by turning to the Soviet bloc in search of weapons and raising the prospect that the canal might fall under Communist control.

Britain and France, fearing an economic catastrophe and also furious at the latest demonstration that their colonial power had crumbled, took an aggressive step—they decided to invade the Suez to protect the canal. Israel, already smarting for a fight to topple Nasser, volunteered to come along for the ride. The only ally they neglected to tell was the U.S., which could only look aghast, not so much at the attempt to overthrow Nasser as at the damage caused to Arab diplomacy by Western para-troops fighting together with Israeli troops.

The Suez drop was a huge embarrassment for the Euro-peans. Arab oil nations promptly banned all oil shipments to Britain and France and the U.S. also declined to bail them out. The Europeans immediately retreated and with them disappeared their influence in Middle East affairs.

Three years later, Nasser again unsettled the West when he helped engineer a military coup against the British-backed Hashemite royal family in Iraq. The new Arabist regime put pressure on the major oil companies and in 1960 it revoked 99.5 percent of the concession granted to the Iraq Petroleum Company, leaving the Seven Sisters companies with only the three fields it was then producing. The Arab nations were be-ginning to flex their muscles. Oil, they realized, could give them great power if they worked together.

Glut and Cut

By the early 1960s, there was far more oil available on the world market than even the multiplying demand for automobiles and passenger jets could sustain. Compounding this glut, the Soviet Union had restored its oil fields to near full production and was now aggressively selling on the world market. The major companies were forced to discount the price at which they sold their Mideast oil, but they still had to pay the producing nations the official price for crude oil they'd set under the fifty-fifty agreements. This meant the companies had to discount their oil at the pump and were receiving only 40 or even 30 percent of the profits. They were, however, still making a handsome profit thanks to a clandestine scam worked out with the U.S. State Department that allowed the companies to deduct the extra money they paid the producing nations as losses on their U.S. corporate tax bill.

Still, they were not prepared to absorb the depressed oil prices alone. In 1959, BP unilaterally cut the posted price of oil by 10 percent, outraging the oil exporters. Two men in particular—Juan Pablo Perez Alfonso of Venezuela and Saudi oil minister Abdullah Tariki—began planning a new oil producers' cartel that would stand up to the Seven Sisters. Then, in 1960, Standard Oil of New Jersey cut the posted price of oil once again, forcing all the other majors to follow suit. This cut into the national economies of the oil-producing nations, all of whom depended on oil profits for the bulk of their national revenue. Within a month, Alfonso and Tariki convened the major exporting countries—Saudi Arabia, Venezuela, Kuwait, Iran, and Iraq—and they agreed to establish a new organization to protect their interests. It would be called the Organization of the Petroleum Exporting Countries, OPEC for short, and it would maintain the price that the producing nations wanted on the global market.

The oil companies realized they'd made a big mistake, but it was still hard for them to believe that this new organization could have any real clout. For the next decade, the seven major companies paid lip service to OPEC while also undermining it by cutting deals with individual member nations.

Still, OPEC's potential and the West's growing dependence on Middle East oil worried the Western powers. Global demand for oil was finally catching up with supply and by 1970, the production glut was over. Unfortunately for the U.S., this rise in demand came just as domestic production reached 11.3 million barrels a day. Never again would it be so productive, and the nation that had come to take cheap oil for granted would from this point on be dependent on foreign oil to meet its needs. The U.S. could no longer count on the security surplus of oil it had maintained for over fifty years. Though it didn't know it yet, America had discovered its Achilles' heel.

By the start of the 1970s, the oil-producing powers could see it was their oil, not that of the United States, that mattered most to the world economy. There had been a twenty-one-million-barrel increase in demand for oil in the non-Soviet world since 1960 and two-thirds of that demand was being met by the Middle East.

Just as the world's oil consumers were dependent on Middle East oil, so were the major oil companies. And that gave OPEC great power whenever it chose to use it.

OPEC's first thrust was spearheaded by Libya. Under the aggressive tack of a new dictator, Col. Muammar Qadhafi, Libya threatened to nationalize all of its oil fields unless foreign companies improved upon the fifty-fifty split. Libyan oil was crucial to the economies of Western Europe—it was easily refined into gasoline and had cheaper transportation costs than Persian Gulf oil—and the companies took his threats very seriously. As soon as Qadhafi upped the old fifty-fifty deal to 55 percent in Libya's favor, Iran demanded a similar deal. Knowing they'd have to offer the same deal to the other Gulf states, the companies agreed to a 55-45 split in the exporters' favor in 1971, along with a 35¢ increase in the posted price of crude oil to be renewed annually.

It didn't last. Within two months, Libya had strong-armed the companies into another price increase and the new agreement would be whittled away by the exporting nations over the next two years.

By 1973, the market price for crude oil had doubled from its 1970 level. But if the Mideast producers were making more money, so were the oil companies. OPEC had intended to reduce the companies' take and redress the balance toward the producing nations, not increase it. Now in October 1973, OPEC announced its intention to set new prices without consulting with Big Oil. The new prices would reflect the desires of the Middle East, not America, Japan, or Europe. As Saudi oil minister Sheik Ahmed Yamani declared in the fall of 1973, "The moment has come. We are masters of our own commodity."

Triple Shock

Just as OPEC ministers were convening to discuss new prices, Egypt and Syria, supported by the Soviet Union, launched a surprise attack on Israel, the start of the Yom Kippur War. Israel, in danger of being overwhelmed, called on the U.S. for help. And when the U.S. military first flew in supplies to its ally (albeit reluctantly) and then okayed a $2.2 billion military aid package for Israel, the Arab nations reacted with indignation.

First, OPEC raised the posted price of oil by 70 percent, bringing it up to $5.11 a barrel, the same price it was currently trading on a now very jittery spot market. Then Saudi Arabia announced its intention to cut off all oil exports to any nation that supported Israel. The other Arab states all did the same.

In the past, the U.S. might have shrugged off this oil blackmail, confident that its powerful Texas, Oklahoma, and California oil fields would simply ramp up production to meet the fall in supply. But those days were over. Saudi Arabia was now the only nation with enough excess producing capacity to act as the "swing" producer—upping production to match a shortfall elsewhere in the world. And Saudi Arabia was the one causing the shortfall.

The embargo, combined with OPEC's price hike, caused panic around the world. By November, oil prices had jumped from $5 to $16 a barrel. At the time, the Nixon administration was so distraught over the embargo that it drew up plans to send troops to the Middle East to seize oil fields in

Saudi Arabia, Kuwait, and Oman. According to British intelligence documents, that move would have necessitated the U.S. to hold onto the oil fields for up to ten years in order to maintain its energy security.

In the end, the Arab states lifted the embargo on Western Europe after those countries pledged their support for the Arab position but waited until the spring of 1974 to resume shipments to the U.S. OPEC had shown that the Arab states finally controlled oil prices and now the main Middle East members began nationalizing their oil industries. The Seven Sisters would find themselves frozen out of Saudi Arabia, Iraq, and, with the 1979 overthrow of the Shah, Iran as well.

The Ford administration took immediate action. In 1975, it established the Strategic Petroleum Reserve, an emergency stockpile of up to one billion barrels of oil that would be stored in empty salt caverns underneath the Texas and Louisiana coast and could be accessed in times of an energy emergency. Today the reserve holds nearly 700 million barrels of oil—just over one month's supply of America's total daily oil consumption. The same year, Congress passed the Energy Policy and Conservation Act, establishing the first-ever fuel economy standards for U.S. cars and trucks.

Western consumers suffered a second oil shock when the Shah of Iran was toppled by the Islamic revolution of Ayatollah Khomeini. Shortly after, the U.S. embassy in Tehran was stormed, fifty hostages taken and all foreign oil companies were thrown out of Iran. This time, Saudi Arabia upped its own production to meet the shortfall but panic once again gripped the world oil markets. Crude oil prices shot up to over $30 a barrel.

Just when it looked like things couldn't get worse, the Iraq regime of Saddam Hussein invaded Iran in 1980, sparking a war involving two of the world's most important oil producers and triggering the third oil shock. Crude prices went over $30 a barrel once again.

Creating an Islamic Monster

The fall of the Shah and the expulsion of U.S. oil companies from Iran had severely rattled the government of Jimmy Carter. Compounding America's energy and security concerns, the Soviet Union had recently invaded Afghanistan in what many observers believed was a blatant first step in a push to control some part of the Persian Gulf. The U.S. believed Iran, with its access to the Persian Gulf, could be the next target for the Soviet Union.

In his 1980 State of the Union speech, Democratic President Jimmy Carter confronted this threat. In what became known as the Carter Doctrine, he said, "Let our position be absolutely clear. An attempt by any outside force to gain control of the Persian Gulf region will be regarded as an assault on the vital interests of the United States of America and such an assault will be repelled by any means necessary, including military force."

To back up the Carter Doctrine, the National Security Council, headed by National Security Adviser Zbigniew

Brzezinski, created a new military unit that could be sent any-where in the world at short notice. It was called the Rapid Deployment Joint Task Force and though originally conceived as a lighting force to assure the unimpeded flow of Persian Gulf oil, it soon evolved into an umbrella command for all of the Persian Gulf, Central Asia, and North Africa. In time it would be renamed the U.S. Central Command.

The Carter administration also helped unleash another force that over the next decade would play a more profound role in Middle East politics than the U.S. military. In a series of secret legal documents known as Presidential Findings, Carter autho-rized the Central Intelligence Agency to begin covert action against the Soviet Army in Afghanistan. Carter could have had no idea just what he had set in motion but, for the best part of the next decade through two successive Reagan admin-istrations, the CIA helped turn a ragtag group of Afghanistan resistance fighters into a well-trained and very well armed mujahadeen—or God's army.

It was a large and costly operation and to help fund it, the U.S. turned to Saudi Arabia for help. The Saudi kingdom had already begun to transform itself into a forward base for the Rapid Deployment Joint Task Force. Convinced that the Soviets had designs on their oil supplies, the Saudis agreed to match the $15 million Congress had appropriated for the Afghan struggle in 1983. In time the budget would increase to $250 million and the Saudis matched it each time.

Thousands of young men from all over the Middle East made the journey to Afghanistan to take part in the jihad—or holy war—against the Soviet infidel. Not only did they succeed in driving out the Soviets but they emboldened the cause of Islam and the growth of Islamic fundamentalism. Much of this new fighting force would return to the Middle East and they would hold onto the weapons—notably the AK-47s, rocket-propelled grenades (RPGs), and Stinger antiaircraft missiles—with which the CIA had supplied them. Before too long, this fighting force would turn against a new infidel—its old benefactor, the United States.

While the Soviets were suffering a slow Vietnam-style humiliation in Afghanistan, the Reagan administration continued to look after its interests in the Middle East. The U.S. still maintained its good relations with Saudi Arabia but it was also eager to support the avowed enemy of its avowed enemy, Iran. His name was Saddam Hussein, and to the U.S. his secular strongman regime in Iraq was a refreshing and stable counterpoint to the Islamic crazies next door in Tehran.

The Reagan administration dallied with Saddam Hussein for nearly three years, but by then, the Middle East had been replaced in importance by the administration's anticommunist campaigns in Europe and Central America. One reason for this was that OPEC members had undercut themselves by overproducing oil and world prices had plummeted. The major consuming nations, meanwhile, began to cut their dependence on the Middle East by consuming less oil and turning to new oil sources from the North Sea, Alaska, and Africa.

Not that Middle East oil could be ignored. In the late summer of 1990 Saddam Hussein once again delivered a shock to the world energy status quo when he invaded Kuwait in a grab for that nation's oil fields. Many observers believed the U.S. had not done enough to dampen Hussein's territorial ambitions in the months leading up to the invasion, but once Iraq attacked, President George Bush moved quickly to assemble an international coalition to repulse Iraq. The coalition was formed in the name of restoring Kuwaiti sovereignty but no one was in any doubt that the security of Middle East oil was the real concern.

Saudi Arabia was particularly worried about Saddam Hussein and it allowed U.S. forces to set up staging bases on Saudi soil for a war against Iraq. The efficiency of the U.S. deployment showed just how closely the two old allies understood each other. Since the Soviet threat a decade earlier, Saudi Arabia had built vast underground warehouses where U.S. weapons and matériel could be stored as well as enormous aircraft hangars just in case the U.S. needed to fly in.

The coalition forces were successful in expelling Iraq from Kuwait but the U.S. left large numbers of troops in Saudi Arabia after the war to deter further Iraqi aggression and to guarantee the security of world oil. In doing so, the U.S. once again stirred Arab resentment of the West and its historic meddling in the Middle East. Muslim anger at U.S. forces posted in the land of Islam's most sacred sites only grew with the return of thousands of well-armed fundamentalist mujahadeen from Afghanistan. One such religious warrior was the Saudi billionaire Osama bin Laden. He declared a holy war on the U.S. for defiling Islam and for its continued support of

Israel. As part of his new jihad, bin Laden would target U.S. interests in Saudi Arabia and Africa. Then, on September 11, 2001 he brought his holy war home to America. Fifteen of the Al Qaeda attackers that day came from America's long-time oil ally, Saudi Arabia.

U.S. forces have pulled out of Saudi Arabia but they now find themselves a de facto occupying force next door in Iraq. The second Gulf War, which drove Saddam Hussein out of power in March 2003, once again placed the United States front and center in the complicated world of Middle East affairs. It is a region that will only grow in importance as the world's dependence on oil grows. And that means U.S. forces will remain in the Middle East for a long time to come.

2

Car Culture:
America's Automobile Addiction

It was a warm afternoon under the hot lights of the New York Auto Show and the crowd squeezed around the new Ford Mustang GT was making it feel even hotter. Nevertheless, the bleached-blond model in charge of promoting the car was intent on stoking the crowd a little more.

"I know you guys love power out there," she said in a husky-gravel voice that made her skintight black leather outfit seem almost sophisticated by comparison. "Well, this supercharged four-point-six-liter engine would blow that young man's hair off," she said pointing to a spotty-faced teenager who turned Ferrari red with embarrassment or testosterone excitement—it was hard to tell.

It was a similar story at other stands. Tough young guys and their girlfriends, proud dads and their sons, and office workers on their lunch breaks all oohed and aahed at the Maybach 57 with a new V-12 and the Cadillac concept car that had been stacked with a sixteen-liter engine. "Go to Cadillac's Web site and vote for this baby to be made," exhorted Cadillac's point man. Even the Pontiac Vibe, a sedate small car by most standards, was being released in a new GTO 5.7-liter version.

Almost unnoticed at the sides of these new displays of Detroit muscle sat the new Toyota Prius whose only number of note was the fifty miles per gallon it gets while being driven in the city.

A Question of Culture

You might have thought that residents of New York City—surely the one place in America where automobiles are more a hindrance than a help—wouldn't go quite so gaga over a car show.

But New Yorkers, myself included, are just like the rest of America—we are under the spell of car culture.

I lived for seven years in Manhattan without a car. I didn't need one—there was a perfectly good mass transit system and over 12,000 taxis to choose from—but not a week passed that I didn't crave my own car. I wanted the freedom a car gives you, the ability to escape New York City whenever I wanted.

Does it make any sense to own a car in Manhattan? Of course not, unless you have the patience and time to cope with the nuances of alternate-side-of-the-street parking. Or unless you are willing to pay the $300-plus monthly fee charged by most New York parking garages. Some city residents pay more to house their cars than themselves.

You'd at least think that, if New Yorkers are so insistent on owning an automobile, they would be clamoring for some compact city-friendly vehicle like those tiny Fiats or Renaults you see zipping through the streets of Rome and Paris. Instead, New Yorkers increasingly crave oversize SUVs that they like to customize with front grille guards—I suppose they think the guards will fend off crazed bike messengers.

SUVs are the most popular new vehicles in America. One out of every four new cars bought in 2003 was an SUV. But they would still appear to be a strange choice for New York City drivers. Most of the driving lanes are too narrow for them to navigate without spilling into someone else's path (though admittedly taxi drivers and delivery vans carve up the streets in exactly the same manner). And SUVs are too big for many parking spaces (and given New Yorkers' bump-and-grind approach to parallel parking, they tend to leave nasty scrapes on other vehicles they've nudged up behind). And parts of the city aren't equipped to handle them. According to New York's chief bridge engineer, a traffic jam of full-size SUVs on the Brooklyn Bridge could cause this architectural icon to collapse. After all, trucks weighing more than three tons are banned from the bridge because of structural concerns. Most large SUVs carrying four adults easily outweigh that.

But SUVs provide New Yorkers with that "don't mess with me" attitude they value above all else. They promise a sense of strength and security that lowly cars can't offer. That's an attribute, however debatable it might be, not to be underestimated in a city still coming to terms with the terrorist attacks of September 11, 2001. As the vamp at the car show put it—everyone loves horsepower.

The automobile is the nexus (and often the Lexus) of daily life in the United States. Americans own more vehicles and spend more time driving them than any other people on earth. There are some 520 million automobiles in the world and the U.S. has 200 million of them, more than there are drivers in the U.S.—1.9 per household compared to just 1.75 drivers per household. In fact, the car population in America is increasing five times as fast as the human population. On average, American drivers cover 11,600 miles a year and spend seventy minutes each day in their car. Most depend on their cars to drive to work, to shop for food, to take their kids to school. American drivers eat breakfast in their cars, conduct business on their cellphones (hands free, of course), and are probably more likely to buy a vehicle based on the quality of its DVD entertainment system than its ability to get thirty miles per gallon on the highway.

Owning a car is a fundamental rite of passage in America, whether you're a teenager brandishing your first driver's license or a new immigrant eager to start up your personal American dream. Automakers and car dealerships make sure cars stay front and center in American minds by spending over $10 billion a year on advertising. Having a car is a way of assimilating in America and, more often than not, it's also the only way of getting around. Public transit systems are rarely found outside of the cities and they carry with them the stigma of catering to those people who can't afford a car. With the automobile being such a necessity and also a symbol of status, is it any wonder that America views life through a car windshield?

Yet this cornerstone of American life is also the main reason for America's dependence on oil. The average new car in the U.S. gets just 23.9 miles per gallon and America's burgeoning fleet of SUVs and light trucks rarely records more than 20 mpg. Oil demand in the United States is projected to grow by 1.7 percent per year to 29.2 million barrels per day in 2025 and most of the increased demand will come from America's drivers.

The Horseless Carriage

Oil and automobiles fell into each other's arms. When Gottlieb Daimler introduced the first gasoline-powered horseless carriage in 1885, the oil industry—including Standard Oil—was reeling from a body blow. Three years before, Thomas Edison had introduced electric-powered light to New York and made kerosene, the oil industry's raison d'être, essentially redundant.

Salvation for the oil industry would come from another new invention and, ironically, it would have one of Edison's own employees to thank for it. Henry Ford was an engineer at Edison Illuminating Company in Detroit who was obsessed with the new horseless carriage, or automobile, as it was known. In 1896 he had succeeded in constructing a gasoline-powered quadricycle and that prompted him to think of building a bigger and better automobile.

While Ford was tinkering with his invention, the early automobiles were embarking on a bumpy road to public acceptance. In Europe, the first automobile race from Paris to Bordeaux and back again had won plaudits from the public but when Narrangansett, Rhode Island, hosted its own auto race in 1896, it was so tedious that one spectator cried "Get a horse" in frustration. Motor cars were slow, smelly, and uncomfortable, but they also were considered very sophisticated by America's upper class. By 1900, there were a few thousand on America's roads and automobile inventors were experimenting with steam-driven and electric vehicles.

Henry Ford, however, was a firm believer in gasoline and in 1908, five years after forming the Ford Motor Company, Ford and his coterie of engineers (Ford's master talent was salesmanship and leadership rather than engineering) unveiled his first gasoline-powered motorcar, the Model T.

It changed everything—the auto industry, the oil industry, the United States and, in turn, the world. Not only was the Model T a reliable, well-designed motorcar—it boasted a powerful four-cylinder engine and a wheelbase that cleared America's rutted roads—but it democratized driving in America. Until the Model T arrived, owning a car had been a luxury. Ford manufactured a motorcar for the people. More important, he devised a way of mass-producing it.

Ford's genius wasn't the car, it was the moving assembly line that produced it. In 1908, the first Model T's cost $850; by 1913, when the first assembly line was up and running at his new Highland Park factory, a new Model T was rolling off the line every three minutes, the cost of the car had halved and Ford had sold almost a million. From 1913 to 1923, production doubled every year and the price of a Tin Lizzie, or "flivver" as the Model T was affectionately known, was cut by two-thirds. The assembly line was a decidely mixed blessing. It offered a motorcar for the masses but it delivered a blow to American craftsmanship and it would take a huge toll on the collective morale of the American workforce. Still, thousands came to work at Ford, not least because he offered a $5-a-day wage, double what other companies were offering.

Ford's thinking was simple. If he paid his workers more, they would be able to buy his motorcars. Model T sales hit 820,445 in 1919, more than all the other cars sold in the country and Ford was spitting them off the production line at his enormous new Rouge River factory at the rate of 120 new cars each hour. By 1923, Ford Motor Company was valued at over $1 billion and Ford's personal fortune at over $750 million.

But just like his former employer Edison, Ford had achieved something far more profound. Both men, with their different inventions, had radically changed the way the world worked. Edison had banished the night and with it the boundaries of time and light that dictated when people should work and rest. Ford, with the assembly line and the motorcar, dramatically increased the speed at which we live our lives.

Registered automobiles grew from 8,000 to 469,000 in the first decade of the twentieth century. And for social critics of America's increasingly overburdened cities, the car couldn't have arrived at a better time. In the eyes of the new progressive movement, Ford's car for the masses was a godsend—it would provide a bridge from the city to the salvation of the country. It would help alleviate overcrowding and the poor sanitation, crime, and disease that accompanied it. And it would replace the fourteen million horses that caused such pollution with their manure.

The automobile was technologically superior to anything that had come before, it was clean in comparison with current industrial know-how, and it captured the optimism of the new century. Furthermore, it appealed to the strong individualism of new Americans. Most had arrived from countries where, because of politics or economics, they had little say in how they lived. Here in the new world they wanted to celebrate a new freedom and they wanted to do it in their own way. That meant driving their own vehicle.

America quickly fell in love with mobility and the road trip. Mass transit—especially the railroads and in the cities the trolley car—suffered at the hands of the automobile. "Gregarious transportation has, of course, made wonderful strides," wrote *American Motorist,* one of the many publications that sprang up around the driving rage, "but the individual has had to sacrifice much of his liberty of action to take advantage of it. He must go with the crowd at the time the crowd wants to go and by the route the crowd takes." If America demonstrated anything, it was that it didn't just go with the crowd.

Carmakers marketed their vehicles as perfect for exploring America's great outdoors and progressives embraced their newfound ability to experience nature (just as many environmentalists would at first embrace SUVs in the 1970s). As early as 1903, city folk were discovering the national park system and by 1922, two-thirds of all visitors to Yellowstone National Park arrived by car and visits to the parks system as a whole grew ten times. Few people considered that the automobile might actually be harmful to the countryside.

By 1920, some two million new cars were being sold each year and America's passion for driving mirrored its post–World War exuberance—*Fordismus* is how the Germans referred to the American giddiness in what we know as the roaring twenties. At the time, local, state, and the federal government were all making sure the automobile would have the road infrastructure it needed to succeed. Much as the introduction of Baron Haussmann's boulevards had brought a sense of space and order to nineteenth-century Paris (while conveniently demolishing the cramped city streets that had been perfect for erecting barricades), so advocates of America's City Beautiful movement envisioned wide new paved roads as the way to bring structure and elegance to the overcrowded cities. But as we now know, wider roads simply begat more automobiles and a new form of city congestion. Not that cars were blamed. It was the trolley car with its fixed route and schedule that took the fall. Mass transit, it was agreed, created gridlock and crushed the free spirit of automotive travel.

The motorcar's charm lay in the opportunities it gave Americans to explore the nation. Urban planners soon began planning new, elegant thoroughfares to sweep motorists out of the cities. Los Angeles's Arroyo Seco Parkway and the Grand

Concourse in the Bronx were two early forerunners of the modern highway. In 1916, President Woodrow Wilson signed the Federal-Aid Highway Act, providing $75 million over five years for a highway department in every U.S. state. There were over three and a half million cars on the road by now and the act was supported both by farm groups and, tellingly, a growing movement of surburban realtors. The push to flee the cities had begun.

Through the 1920s, new highway investment across the country reached a billion dollars a year and provided the foundation for the automobile and domestic oil industry. The four-lane Bronx River Parkway opened in 1925 and asphalt fever soon captured the imagination of engineers and city planners. At the vanguard of this movement was Robert Moses. As parks commissioner for New York City, and later as near-omnipotent head of the Metropolitan Transit Authority, he would create elegant asphalt tentacles out of the city in the shape of the Saw Mill River Parkway and the Hutchinson River Parkway before later forcing his autocratic auto vision on the city itself, leveling whole neighborhoods to build the Brooklyn-Queens and Cross Bronx Expressways.

The new roads made living outside the city attractive. Not only was the price of real estate cheaper on the edges of the city, but there was, of course, more space—not least to park one's car. The notion of city suburb had been pioneered by the streetcar, which had been quietly expanding the boundaries of metropolitan living since the turn of the century. Now the motorcar blew the idea of city limits apart. The 1920s saw a 59 percent growth in suburban populations as new auto-inspired communities like Grosse Point outside of Detroit and Elmwood Park near Chicago flourished.

The investment in creating a new mobile American way of life would come to a halt with the 1929 stock market collapse and the onset of the Great Depression. But that didn't mean the cars stopped running. "America was the only nation in the world that ever went to the poorhouse in an automobile," Will Rogers joked at the time, and somehow the mass unemployed, like John Steinbeck's plucky Joad family in *The Grapes of Wrath,* always found a little something extra to keep their cars on the road. Standard Oil even began selling a new bottom-of-the-barrel "fighting" brand of gasoline, Stanolind Blue, for just 12¢ cents a gallon to keep America on the move. The reason was simple—the car had not just increased mobility for sightseeing, it greatly expanded the chance of finding a job. From now on, U.S. workers would think nothing of uprooting family and moving sometimes thousands of miles across the country for new work. This new American fluidity would be an enormous asset to the U.S. economy in the years immediately following World War II, when a ready and eager mobile workforce decamped from the Northeast to the Southwest in search of new opportunities.

Just as the United States was at its lowest ebb, the automobile received its biggest push. In 1932, Franklin Delano Roosevelt promised his New Deal, a centerpiece of which was the construction of new roads. As part of the emergency Public Works Administration initiatives, one million people were hired for highway projects. FDR's Civic Works Administration allowed for 500,000 miles of road building and some 80 percent of all New Deal agency expenditures were dedicated to roads and new construction.

The full implications wouldn't be realized for nearly fifteen years when a generation of young soldiers, sailors, and airmen returned from World War II. They found inspiration in the vision of the Levitt family, who in 1947 transformed 4,000 acres of potato farming land on Long Island, twenty-five miles east of New York City, into affordable grouped housing. The houses cost no more than $9,500, making them affordable to middle-class folk seeking to escape the city. Isolated from New York's public transit system, the easiest way for these new residents to visit their families back in Brooklyn and Manhattan was by car. The second wave of suburbanization had begun and this time there would be no Depression to curtail it. Nine million people moved to the suburbs between 1945 and 1950 and this exodus would shape the future of urban life in America. Another eighty-five million people would join them between 1950 and 1976 while the major cities would add only ten million people during the same time.

Automobile hegemony was already assured in society when the motorcar received a final defining boost by one of America's first road trippers. President Dwight D. Eisenhower was just a young army captain when, in 1919, he volunteered to take part in a drive cross-country from Washington, D.C., to San Francisco to celebrate the potential of motor transport. The trip took a torturous two months over roads that broke axles, fan belts, brakes, and every other possible moving part. And yet Eisenhower emerged from the experience convinced of the value of the automobile and, equally important, of the need for a decent road network. In 1956, during the end of his first term in office, he signed the Interstate Highway and Defense Act. It would enable a 41,000-mile highway system to connect the whole country and the federal government would cover 90 percent of the cost. The project, he boldly announced, was necessary to ease congestion, improve safety, and ensure national security by providing easy movement of U.S. nuclear missiles in the face of a Soviet nuclear attack. It would "change the face of America," Eisenhower said. He had no idea just how much.

Tiger in the Tank

In the early years of the motorcar, refueling your car meant carrying extra gasoline with you. But in 1905, Harry Grenner and Clem Laessing formed the Automobile Gasoline Company in St. Louis and began selling gasoline to drivers using a gravity-fed tank attached to an ordinary garden hose. Soon, the two men had opened a chain of forty such "gas stations" throughout St. Louis. In Seattle, John McLean, the area sales manager for Standard Oil of California (Socal), went one better, opening the first drive-in "filling station," as he called it, in 1907. When McLean's customers pulled in for gas, they could also purchase Standard's Zerolene engine grease and Polarine motor oil.

The oil industry was taking the first steps to transform its amorphous product into a sexy consumer brand that the public could identify with. Yet gasoline was not going to drive America's cars or its economy unless the oil companies could produce more of it. For the previous forty years, companies had focused on getting the most amount of kerosene from crude oil. Gasoline was a by-product of the refining process, and one barrel of oil yielded at most 20 percent gasoline.

One year after Ford's Model T went on sale, a Standard Oil executive, anticipating the need for more gasoline, began fashioning a solution. William Burton was head of manufacturing at Standard of Indiana and had a Ph.D. in chemistry. His aim was to crack—or break down—the larger molecules contained in fuel oil by heating them under high pressure up to 650 degrees to produce more gasoline. By 1913, his first thermal cracking still, producing 45 percent gasoline from a barrel of crude, went into operation just as gasoline sales in the U.S. overtook those of kerosene.

Burton's invention ensured there would be enough gasoline to satisfy the car-crazy public. But that still left the question of how the oil companies would get the gasoline to the consumer. By 1914, Socal had thirty-four standardized filling stations in California, but most gasoline was still sold by mom-and-pop general stores and other miscellaneous businesses. The oil companies continued to recruit small businesses as new outlets for selling gasoline because it meant they didn't have to invest in new buildings and could reap a quick profit. Soon, gasoline pumps were cropping up outside any curbside business that

REFINING: HOW CRUDE OIL BECOMES GASOLINE

Carbon is one of the few elements that create extended chains or elemental branches. So while all crude oil is a combination of just hydrogen and carbon, no one crude oil is identical to another. Crude oil is of little use by itself, but, when heated, its various components can be collected as they vaporize at different temperatures. This is how crude oil is refined.

Oil's various refined products have progressively higher boiling points. The lighter the chemical compound of the product, the lower the temperature needed to separate it from the crude oil. First to be refined are methane, ethane, propane, and butane gases. Next come clear liquids called naphthas that are used as solvents and dry-cleaning fluids. Then come a series of hydrocarbon chains that vaporize below the boiling point of water and are blended together to form gasoline. After gasoline comes kerosene and then lubricating oils such as motor oil and vasoline. At the end of the refining process a thick residue remains. It is used to make asphalt.

Refineries restructure the crude oil molecules through a process called cracking. Lubricating oils and kerosene can both be cracked to produce more gasoline. Refineries are already so proficient that they can actually extract 44.6 gallons of refined petroleum products from a 42-gallon barrel of crude. The extra yield is called a "processing gain."

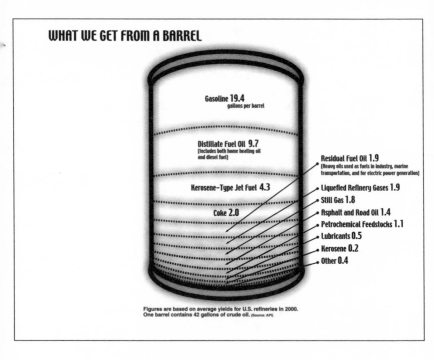

WHAT WE GET FROM A BARREL

Gasoline **19.4**
gallons per barrel

Distillate Fuel Oil **9.7**
(Includes both home heating oil
and diesel fuel)

Kerosene-Type Jet Fuel **4.3**

Coke **2.0**

Residual Fuel Oil **1.9**
(Heavy oils used as fuels in industry, marine
transportation, and for electric power generation)

Liquefied Refinery Gases **1.9**

Still Gas **1.8**

Asphalt and Road Oil **1.4**

Petrochemical Feedstocks **1.1**

Lubricants **0.5**

Kerosene **0.2**

Other **0.4**

Figures are based on average yields for U.S. refineries in 2000.
One barrel contains 42 gallons of crude oil. (Source: API)

wanted them, with little regulation on the quality of gasoline sold or how safely it was dispensed. It took a series of gas fires and explosions at these often overcrowded and poorly maintained pumps to prompt local government to begin zoning where gasoline could be sold and what structure was required to house the gasoline tanks.

This gasoline free-for-all was also hurting the oil companies' brand image and that was increasingly important as oil prices plummeted and the companies had to compete harder against each other. Their solution was to construct stand-alone gasoline emporiums—service stations, as they were called east of the Rockies, and filling stations, as McClean had coined them, to the west. Some 143,000 drive-in gas stations had been built by 1929 and, to identify and market their brands, the companies devised logos that they hung high throughout the nation for passing motorists to see. The companies also aggressively marketed new types of "improved gasoline" that promised better performance than their competitors'; some even colored their gas with blue and red dyes to help build brand recognition. Drivers quickly came to identify with the brand they could trust, be it the star of Texaco, the dinosaur of Sinclair, Shell's scallop shell, or Socony's Pegasus. This loyalty might remain for years and be passed down through the family from father to son. Driving, after all, was still very much a male domain.

The oil companies were also selling something more intoxicating than the fumes of gasoline. They offered their customers a sense of freedom they had never experienced before. And, with so many of the new middle class driving the same Model T car, the different gasolines offered a whiff of individuality, a far cry from the communal drudge of the Old World.

In 1928 advertising icon Bruce Barton delivered a messianic speech to a gathering of top oil executives. Brand awareness was already a buzzword through the industry but Barton wanted his audience to understand something higher—the "magic of gasoline."

"It is the juice of the fountain of youth that you are selling," he told them. "It is health. It is comfort. It is success. And you have sold merely a bad smelling liquid at so many cents per gallon. . . . There is a magnificent place for imagination in your business, but you must get it on the other side of the pump. You must put yourself in the place of the man and woman in whose lives your gasoline has worked miracles."

Motorists pulling into one of the big-oil-company gas stations during this time could expect better service than if they sat down to eat at a fancy restaurant. Not only was gas dispensed by a snappily dressed attendant but now tires were checked, the windshield wiped clean, and the radiator topped off with coolant. The companies started outfitting their employees in uniforms—Shell Oil attendants, for instance, wore jodhpurs, a matching waist-length jacket, a bow tie, and shiny leather shoes.

The companies also started giving away gifts to lure customers. Free maps had become an extra offered by gas stations starting in the 1920s and in the winter of 1929, Shell decked out some of its stations with bright Christmas-light displays. The Phillips Petroleum Company started giving away toys to children when it opened a new station and by the 1950s, Socony had mastered the art of winning over its future customers by offering mechanical Pegasus horse rides for children at its own new station openings.

The new gas stations gave automobile drivers a sense of familiarity and dependability as they traversed the nation, but what were the drivers to eat? And where would they stay while on the road?

Southern California led the way. The first motel opened in 1926 in San Luis Obispo charging $2.50 a night. California had been attracting newly mobile migrants from the East since the early days of the twentieth century. More people were drawn from the dust bowl states in the Depression years and by the early 1940s, hundreds of thousands more arrived in search of the high-paying jobs now available. Between 1940 and 1945, the federal government invested $20 million on the West coast to build aircraft factories, shipyards, and military bases far removed from the threat of a German attack.

Some two million people moved to southern California between 1920 and 1940 and many would choose its main city, Los Angeles, as home. Unlike the older cities back east, L.A. had no entrenched transportation infrastructure, and so the automobile would determine the shape and breadth of the city. L.A.'s trolley system was dismantled when it was secretly purchased by General Motors, along with similar networks across the nation. The carmaker also persuaded Socal, Mack

Truck, and Firestone to take part in the conspiracy and all were found guilty of antitrust violations for gutting L.A. mass transit in 1947.

By then the automobile had won—California was a driving state with L.A. claiming over a million vehicles to its name. And for the state's burgeoning and restless teenage population, nothing was more important than the drive-in restaurant.

"People with cars are so lazy that they don't want to get out of them to eat," was how one early drive-in restaurant chain owner put it as recounted by Eric Schlosser in his book *Fast Food Nation*. Drive-ins gave young men a place to hang out, to play music on new jukeboxes, to meet girls or just check out the long-legged waitresses—carhops, as they were called—and, oh yes, to order some food as well.

The drive-in's runaway success would be short-lived, for it was soon to be overshadowed by a far simpler concept—the drive-through. In 1948, Richard and Maurice McDonald were running a small drive-in restaurant in San Bernadino, about sixty miles east of L.A. But, tired of the grind of running a regular restaurant and tired of the boisterous and pilfering teenage clientele, they decided to strip down their business by copying the factory assembly line.

The McDonald brothers reduced their menu by two-thirds and focused almost solely on hamburgers. They did away with cutlery and they did away with the carhops, making their customers fetch their own food. But most important, they took a leaf out of Henry Ford's book and started preparing their food in an assembly line. One person would cook the burger, another prepare the bun and condiments, another cook the fries, and yet another make the shakes. The result was culinary economies of scale and cheap fast food for the customers. California's young families loved it but McDonald's might have remained merely one state's secret had it not been for the McDonald's milkshake blender vendor, an Illinois salesman named Ray Kroc, who persuaded the brothers to let him franchise the McDonald's name and system nationwide.

While suburban America, epitomized by southern California's sprawl, expanded, America's inner cities crumbled. First out of the door were the retail stores that could no longer count on their customers to venture downtown to shop. In 1949, the first shopping center targeted at automobile drivers opened in Raleigh, North Carolina. Soon, suburban malls were flourishing and sending urban mom-and-pop-owned stores into early retirement.

The Age of Extravagance

Cheap, plentiful oil supplies along with postwar euphoria and newfound spending power drove the consumer cultural revolution that swept through the U.S. in the 1950s and early 1960s. The theme of the age was bigger is better and Detroit embraced it wholeheartedly, turning out such classics as the Cadillac Eldorado Brougham in 1957, part of a new trend of flashy celebratory automobiles.

During this period Europe also fell in love with the sleek, long, and elegant American automobiles they saw each week at the movies. Still, despite their infatuation, Europeans built much smaller cars than the United States. One reason was geographical—people in France, Germany, and the UK didn't need to cover the same distances as American drivers—but the main reason was economic. Since the 1930s, European nations had been heavily taxing gasoline. So few people had cars at that time, they were considered a luxury good. European governments were also acutely aware that they did not possess any substantial oil reserves. The U.S., on the other hand, was the largest supplier of oil in the world and its well-established domestic producers had long fought increased taxation on gasoline. So while European cars remained for the most part compact, American vehicles grew to outlandish proportions and tended to be powered by V-8 and even V-12 engines that got little more than ten miles to the gallon.

That was fine while the U.S. and the major Western oil countries could control the price of world oil. But U.S. oil production peaked in 1970 and OPEC began dramatically hiking the price of crude oil. Oil prices would rise from $5 a barrel in 1973 to nearly $35 a barrel by 1981, and driving a muscle car became a luxury few Americans could afford.

Congress passed the Energy Policy and Conservation Act in 1975 that established the Corporate Average Fuel Economy (CAFE) standards for passenger cars. It raised the average miles-per-gallon performance of U.S. cars from 13 mpg in 1975 to 27.5 mpg by the 1984 model year. The new rules looked at a company's fleet of cars as a whole, but because most of the

Big Three's vehicles were big, unwieldy boats they had little time to redesign these large cars to make them more fuel efficient.

Detroit rushed a new generation of hastily designed small cars into production like the infamous Chrysler K car series. These new cars helped the automakers meet fleet CAFE standards but they were a mechanical disaster and motorists hated them. The new cars were lower to the ground than before and had less headroom, making them harder for older people to drive. Some of the new engines had been so quickly redesigned that they were forever breaking down. Worst of all, they were not as safe as the older, heavier American car fleet. To reduce weight and hence improve fuel efficiency, the Big Three had basically just shrunk the front and back ends of older models and cut down on the framework to save weight.

Perhaps if Detroit's Big Three had had more time to think about a small car fleet they might have been able to woo American motorists away from their love affair with large automobiles. But in the eyes of most drivers, the small American cars that were produced in the 1980s represented a heresy to the great American automobile tradition. The majority of motorists who still preferred U.S. cars over Japanese or European imports equated the high gas prices with underpowered, badly made small cars. And by the time OPEC pumped its way into an oil glut in the mid-1980s and oil prices dropped below $10 a barrel, many drivers were more than ready to return to a big, sturdy passenger vehicle.

Bigger and Badder

It would be wrong to blame all of the world's energy woes on America's infatuation with the sports utility vehicle. After all, there are only twenty million SUVs on U.S. roads. In the years to come, as car ownership increases all over the world, other nations will contribute more to global warming than the United States.

But our love affair with SUVs still puts us at a particular disadvantage as we try to rein in our dependence on foreign oil.

In his history of the SUV, *High and Mighty,* Keith Bradsher recounts how a legislative loophole wide enough to drive a Lincoln Navigator through gave birth to SUVs and how they came to play such an important role in the fortunes of Detroit's Big Three automakers. When Congress established CAFE standards in 1975, the auto industry lobbied hard to make sure they would not be applied to light trucks. At the time, those vehicles made up just 20 percent of the market and were used mainly by farmers and other small businesses. The auto industry argued that having light trucks conform to the same environmental standards as cars would hurt the U.S. economy. American companies had cornered what at the time was a niche market and they insisted that the cost of redesigning the trucks would lead to many job cuts.

At the time, the automakers were also experimenting with a new large vehicle based on a light-truck design but targeted very much as a passenger vehicle. Executives at the AMC corporation had been struck by how many urban dwellers were enamored with the classic Jeep truck. AMC redesigned the Jeep so that it was more passenger-friendly and termed it a sports utility vehicle. Then it began selling these SUVs to upscale buyers as a weekend getaway. AMC and other automakers argued that sports utility vehicles should be classified as light trucks in terms of fuel economy standards and Congress acquiesced. In time this would become known as the SUV loophole, an oversight that would see America's oil consumption skyrocket and its automobile fuel-economy standards tumble.

Exempt from CAFE standards for cars, light trucks were subject to much lower fuel-economy levels (the Department of Transportation set an average of 20.5 miles per gallon for light trucks by the 1987 model year) so Detroit had little incentive to invest in upgrading light-truck engine technology. That made the vehicles much cheaper to manufacture than cars but far dirtier and less efficient. For the automakers, whose whole business model had been threatened by the new push for conservation and better fuel-economy standards, SUVs offered the chance to make a lot of money in a market where there was virtually no competition.

Fuel-economy levels for new automobiles peaked at 27.5 miles per gallon in the 1988 model year. That's exactly the time when SUVs began to capture the imagination of the U.S. driving public.

The average SUV burns 40 percent more gasoline than the average car. SUVs can spew up to one and a half as much carbon dioxide as cars, exacerbating their contribution to global

A GLOBAL WARNING

Rising sea levels, violent weather patterns, long-lasting drought conditions, and melting ice caps. These are all signs that the earth is getting hotter. And it is humans, agree most scientists, who are to blame.

This phenomenon of rising temperatures is called global warming. It occurs when carbon dioxide and other heat-trapping gases collect like a blanket in the atmosphere. The gases come from cars, power plants, and other industries that burn gasoline, coal, and other fossil fuels.

1998 was the hottest year ever recorded. 2002 was a close second. The industrial world has raised global temperatures by about one degree Fahrenheit. But most of that change has come since 1970. During that time, the depth of Arctic permafrost has shrunk by 40 percent; Andean glaciers in Peru are retreating; summers are hotter and wetter in Europe; hurricanes and typhoons are more ferocious in the Atlantic and Pacific; and drought conditions in China are getting worse—more than 2,500 kilometers of land turns to desert in China each year. By 2015, the snow-capped peak of Mt. Kilimanjaro will no longer have any snow, and by 2030, Montana's Glacier National Park will have to make do without any glaciers.

Rising sea levels threaten major coastal cities around the world. If the oceans rise by one meter, Bangladesh will lose 17 percent of its landmass. And as water temperatures increase, vast tracts of coral reef—the so-called lungs of the oceans—could be decimated.

Since the beginning of the Industrial Revolution, humans have increased the amount of CO^2 in the earth's atmosphere over a third. New projections suggest that global temperatures will rise by another five degrees Fahrenheit over the next 100 years. That will make the earth hotter than it has been in 400 million years.

warming. And because the existing rules allow car manufacturers to be judged on the average fuel economy of their whole
fleet, companies have increased production of small, light-
weight vehicles to offset the production of the best-selling
heavy SUVs and trucks. This has created a dangerous imbal-
ance in road accidents. While light trucks represent 36 percent of all registered vehicles, they are involved in nearly half
of all fatal two-vehicle crashes with passenger cars. Eighty percent of those killed in those accidents were riding in the cars.

Detroit has always argued that it is simply producing the
type of vehicle Americans want. Critics counter that U.S.
drivers became so intimidated by other SUVs on the road
(and the fear of being hit by one) that they felt the need to
buy a big vehicle too. It's as if some Darwinian theory of automotive evolution took effect.

The irrational exuberance of the 1990s certainly played a
role. At some point after the first Gulf War, as gas prices fell
to historic lows and Americans became more consumed with
checking their dotcom stock portfolios than the everyday
price at the pump, the public forgot the oil lessons of just a
few years before.

Between 1990 and 1997, annual SUV sales jumped from
708,000 to 2,446,000 as Detroit created new luxury models and
foreign automakers also got in on the act. By 2001, they would
hit three and a half million. That same year, sales of large SUVs
rose from 0.6 percent of the U.S. auto market to 7.1 percent.

Celebrities were among the first drivers to embrace the
vehicle. Ever since Arnold Schwarzenegger took to driving

around L.A. in a military Hummer after the first Gulf War, celebrities have elevated the SUV to, well, celebrity status. Luxury SUVs have become as important in rap music videos as near-naked women and Cristal, while movie stars routinely pull up to red-carpet openings in SUVs.

But SUV culture speaks to a deeper American trait than its penchant for celebrity worship. It reflects America's own insecurities in the greater world. The U.S. is the strongest, most dominant empire the world has seen for centuries. Its military, economic and cultural influence touches nearly every person on earth, yet Americans still feel that they are the ones who are picked on.

The Hummer remains a mini-tank. And in the eyes of a nation that has always had a suspicion of the outside world and which now realizes its vulnerability to a foreign attack, the SUV offers a sense of security.

When New York mayor Rudy Giuliani arrived to tour the wreckage of the World Trade Center, he did so in his trademark black Suburban SUV. With America at war, GM's H2, the slimmed-down civilian version of the army Humvee, has been the best-selling luxury SUV at a rate of 3,000 a month since being released in 2003. As one H2 owner told the *New York Times,* "If I could get an A-One Abrams [tank], I would."

But while America is preoccupied with the threat it feels from abroad, it is still far more concerned with the smaller everyday fears of American life—the fear of crime, the fear of getting caught in the rain or the snow, and the fear of getting muscled out of a parking spot at the mall by a bigger SUV.

The auto industry has played on those fears, making imposing-looking SUVs like the Dodge Durango and the Cadillac Escalade and marketing its SUV fleet with such slogans as this one for the Chevy Avalanche: "We didn't intend to make other trucks feel pathetic and inadequate, it just sort of happened."

Mothers have been persuaded that their kids won't be safe in an accident unless they drive an SUV, even though all research shows that SUVs roll over in accidents more than any car, that the increase of SUVs on the road cuts down visibility for all drivers, that SUVs have a worse chance of avoiding an accident due to their size and handling, and that smaller SUVs offer no better protection than cars when slammed at 35 mph by the new monster trucks on the road. Even the automakers have admitted that large SUVs pose a threat to other motorists.

Yet America's drivers keep buying these monstrosities. In a time of such uncertainty, when Americans are told they could come under attack at any moment, climbing into an SUV can still feel like you have entered an impenetrable fortress. And, as long as no one else on the road can hurt you, how you drive, and where you are headed, doesn't really matter.

3

Exploration or Exploitation?: Oil, Human Rights, and the Environment

Guillermo Maldonado was forty-seven years old but he looked nearly sixty. "We have bad headaches, stomach pains, and our skin looks like this," said the Ecuadorian oil day laborer as he turned over his arms and rolled up the sleeves of his work shirt. They were covered, just like the rest of his body, with white pus–filled boils. It could have been a hereditary disease—the rest of his family all suffered from the same condition—except that Maldonado's neighbors also had the same affliction.

Maldonado worked for Petroecuador, Ecuador's state-owned oil company, and he lived in a jungle shack on a dusty trail a few miles outside of Lago Agrio, Ecuador's biggest oil town. Situated on the edge of the Amazon rain forest, Lago Agrio was once considered the sacred land of the Cofan Indians but the Cofans had been forced out years before when thousands of poor highland Indian settlers like Maldonado, lured by the promise of secure work and good money, moved to the region as Texaco started drilling here. They never made a fortune. Instead all they found was poverty and illness.

Texaco, now ChevronTexaco, spent seventeen years from 1971 to 1991 drilling for oil in the Oriente, one of the most ecologically rich and diverse regions in the world. And by all accounts, save those of Texaco, the drilling has left an enormous mess. The company, it has been widely alleged, spilled 16.8 million gallons of oil (one and a half times that of the 1989 *Exxon Valdez* oil spill in Alaska) from ruptured pipelines, discharged 19 billion gallons of highly toxic water into waterways and the soil, conducted massive deforestation, and left more than 600 toxic pits uncovered.

Maldonado walked the short distance from the wooden shack he shared with twelve family members to a toxic water pool first dug by Texaco decades ago and still used by Petroecuador as part of its petroleum separation process. Near the pool, a gas flare burned with an intense heat, releasing invisible fumes that sent my head spinning and made me feel sick to my stomach. The pool was filled with black, slimy water and was lined with a single layer of black plastic. The plastic had ruptured and the water leaked through the soil and into a nearby stream—the only local source of drinking water.

A few miles away, another subsistence farmer, forty-six-year-old Vincente Alban, lived right next to a toxic pit. Alban plunged his machete into the sodden land he used for fifteen years to graze cattle and pulled up a thick clump of black crude. "I've lost between twenty-five and thirty cows during the time I've been here," he said matter-of-factly.

The pit had just been "cleaned" by one of Texaco's sub-contractors, as part of a 1995 agreement by the company with the Ecuadorian government to remedy some of the waste it left behind. But, in most cases, the cleanup consisted of little more than filling the pit with earth, a process that flooded Alban's land with crude oil and toxic water. "All the water in this town is contaminated," said Alban, pointing out that only rainwater was drinkable. And that the local people collected it in the only large containers they could afford—discarded Texaco oil drums.

A Blessing and a Curse

To the average U.S. consumer, oil, if not the companies that exploit it, is a good thing. Oil, after all, has driven the American economy for decades and brought great prosperity to the country. At the same time, the pioneering tales of the Texas, Oklahoma, and California oil strikes remain a testament to the can-do independent spirit that America has always embraced. Yet increasingly, Americans seem wary of what new oil drilling might do to their own communities—of the environmental consequences, and the economic damage oil drilling can do to other industries like ranching and tourism. Even Florida governor Jeb Bush, for example, balked at his brother George's plans to open up Florida's coast to new oil drilling in 2002. The president, realizing a potential public relations disaster in a state he had to win in order to regain the White House in 2004, killed the project.

Environmental pollution, human rights abuses, political murder, and civil war—these are legacies of too many nations that possess oil. From Angola in Africa to Myanmar in Southeast Asia to Ecuador in South America, oil-rich nations are more prone to social and economic conflict than nations that claim no significant natural-resource riches.

There are exceptions, of course. The United States, along with other major oil producers such as Canada, Norway, and the UK, have all prospered from oil development without falling victim to widespread corruption and environmental pollution. But these nations were all established democracies before oil was discovered and all had well-rounded economies which oil could bolster rather than dominate.

Most of the countries that find themselves suddenly endowed with oil wealth don't have the security of a well-established democratic and accountable system of government. And they don't have robust and well-integrated economies that can help cushion the pressures of new quick oil riches. Instead, these countries tend to be economically fragile with weak democracies or autocratic regimes and they all want to make a lot of money quickly. Oil, unfortunately, only encourages autocratic rule and undermines the economies of these countries.

These are constant themes that permeate through all developing nations "blessed" with oil. But in understanding just how destructive oil can be to the environment, the people, and the fabric of a country's society, there is no better example than Ecuador. Through forty years of oil exploration, the people of this small Andean nation have seen their land ravaged and their natural resources plundered without most of them enjoying even the hint of improved prosperity.

When Texaco arrived in Ecuador in the late 1960s, the country was living in the shadow of its neighbors, Colombia to the north and Peru to the south. The capital, Quito, had its faded Spanish Colonial architectural splendor to recommend it but little had been done in the way of public works improvement since the Spanish left. The discovery of impressive new oil fields in the Northern Oriente—as Ecuador's Amazon region is known—changed all that. As foreign oil companies and petrodollars flooded into the country, a new part of Quito, literally called the New City, grew up directly to the north of the rambling old town. This new city was dominated not by churches and plazas but by steel-and-glass highrise office towers that reflected the instant wealth generated by the oil boom. Just in case there was any doubt who was responsible for Ecuador's economic turnaround, many new buildings boasted the corporate logos of the oil companies that were transforming the nation.

For a while in the 1970s, Ecuador saw handsome profits from its oil earnings and the military government invested in more and more public works schemes. But when oil prices plummeted in the early 1980s Ecuador was left mired in debt and without the necessary oil revenues to meet its payments.

In 1970, Ecuador's foreign debt was $217 million. Today, Ecuador is over $13 billion in debt to foreign institutions, including the U.S., and only its profits from oil keep the country afloat. Thirty years after the first well was sunk, most of the $32 billion that the oil industry has created has gone to foreign oil companies or is sitting in the Miami bank accounts of an elite group of Ecuadorian politicians, generals, and businessmen.

For the rest of Ecuador society, life has not improved. Ecuador always had high levels of poverty—the indigenous Andean people and the lowland Amazon tribes had been living a subsistence lifestyle for centuries. But by 1995, according to a World Bank study conducted at that time, more than 70 percent of Ecuadorians were living in poverty and almost half of all children were malnourished. More than six million Ecuadorians had income levels that did not even permit them to satisfy their basic needs. The World Bank estimates per capita income in Ecuador is U.S.$1,360. In a country where life had always been tough, oil had actually made things worse.

The Dutch Disease

In the parlance of academia, Ecuador has "Dutch disease," an almost total reliance on oil revenues that comes at the expense of other non–natural resource industries. In the late 1960s, the Netherlands developed a hard-and-fast addiction to the easy revenues it was collecting from its North Sea natural gas reserves to the detriment of its overall economy. Dutch disease suggests that overwhelming oil wealth undermines national economies—with so much money to be made from oil or gas, there is little incentive to develop competitive manufacturing and service sectors. At the same time, a country whose economy is ruled by oil exports will likely have an exchange rate that favors imports and hinders exports. So while oil remains profitable, other industries see their products become more expensive and uncompetitive abroad.

At least the Netherlands' oil problems started offshore. Ecuador had the added misfortune of finding large deposits of oil slap in the middle of one of the world's most important centers of biodiversity—the Amazon rainforest. In the 1970s, Ecuador's government encouraged thousands of highland residents to resettle in the Oriente region, promising that the new foreign oil companies would give them jobs and rescue them from poverty. And eager to reap the maximum reward for the oil boom, the government encouraged the foreign oil companies to drill and pump as quickly as possible.

Ever since the formation of OPEC first sent international oil companies scurrying out of the Middle East in search of new, more diverse, and less problematic sources of oil, the companies have been aggressively courting any nation with promising new reserves. Oil companies negotiated either joint operating or profit-sharing agreements with the host countries. The scope of these can include drilling and refining in conjunction with the state-owned company or being put in charge of all oil production and just paying "rents" to the country for the right to drill.

In the United States it took the government years to come to terms with the damage that industry was causing both to the environment and to the lives of the people living around it. The American environmental movement cut its teeth on issues like the widespread risks of water pollution that came from the use of DDT in agriculture and the toxic dumping at Love Canal in New York State. In time, tough laws like the Clean Air and Clean Water Act were passed to create a system of checks and balances on the environmental abuses industries often commit in the push to increase profits. Today, even in the U.S., those protections have been undercut by government reforms that propose that industries like oil and coal be trusted to regulate themselves.

So imagine then how oil companies have conducted themselves in countries where environmental and health laws are nowhere near as defined as in the U.S. and in many cases don't exist at all. Throughout the world, the marriage of developing nations desperate to cash in on their newfound riches and oil companies eager to drill and export the oil quickly in order to maximize their own profit has been disastrous for the people and the land where the companies choose to drill. In Ecuador, this rush to plunder was exacerbated by the huge new migration to the oil regions, traditional home of previously sheltered Amazon indigenous tribes.

Texaco's Toxic Legacy

Texaco spent seventeen years pumping oil out of the Oriente. During that time, its production consortium accounted for 88 percent of all the oil shipped out of the region—some 1.4 billion barrels in total. Also during that time, it created

THE SUPERMAJORS

The oil industry remains the biggest business in the world. It is valued at between $2 trillion and $5 trillion (the discrepancy comes from the difference in considering oil as either a commodity or manufacturing good). Either way it dwarfs most other organized economies. In recent years, a spate of mergers has created three new super-companies, ExxonMobil, BP, and Shell.

ExxonMobil

The merger of ExxonMobil in 2000 represented a rejoining of two of the biggest companies in the original Standard Oil Trust—Standard Oil of New Jersey (Exxon) and Standard Oil of New York (Mobil). ExxonMobil is active in over 100 countries and is the largest publicly traded company in the world. In 2000, the company had annual corporate revenues of $210,392 million, which would make it the twenty-first wealthiest nation in the world, larger than the GDP of Turkey, Austria, and Hong Kong.

BP

London-based BP began life in 1909 as the Anglo-Persian oil company. The current company is the result of a merger with Amoco and ARCO. BP possesses proven reserves of 17.6 billion barrels of oil and gas. It produces 2 million barrels of crude oil and 8.7 billion cubic feet of natural gas each day.

Shell

A culmination of the earliest major oil merger in 1907, two companies, Royal Dutch and Shell, together control shares in a group of companies that operates in 145 countries and employs more than 115,000 people. Shell is the world's leading private producer of liquid natural gas. The company recorded earnings of $9.2 billion in 2002.

the havoc that ruined the land new immigrants like Vincente Alban and Guillermo Maldonado had settled on. Texaco's 312-mile pipeline over the Andes to the Pacific coast had ruptured at least twenty-seven times, spilling 16.8 million gallons of crude into the region's delicate environmental framework. And the company had dumped toxic wastewater from its operations into open pits, lined with just a single layer of plastic. These pits soon ruptured, allowing a combination of oil and toxic water to leach into the ground and nearby aquifers, infecting the drinking water of the communities living close to the company's drilling operations.

When Texaco left, it handed over operations to Petroecuador, which continued the same practices Texaco had begun in the early 1970s. When New York researcher Judith Kimerling visited the Oriente in 1989, she was horrified to learn that, not only had none of the old spills and toxic pits been cleaned up, but the petroleum industry was still dumping 4.3 million gallons of untreated wastewater into the local watershed every day. Kimerling documented her findings and they would later be published as a book, *Amazon Crude.* It would be an important first step in helping Ecuador's local communities stand up to oil exploitation.

In 1993, a lawsuit was filed on behalf of 30,000 indigenous rain forest people alleging that Texaco knowingly polluted the land. The case has bounced back and forth between U.S. and Ecuador courts ever since and in the fall of 2003, it finally came to trial in the oil town of Lago Agrio—where Texaco began its operations years before. Texaco no longer has any operations in Ecuador, but the U.S. courts have indicated they will enforce any financial judgment against the company that results from this trial.

At the trial, lawyers for the plaintiff presented a Texaco internal document that indicated the company had rejected lining its earthen waste pits to protect the environment because it would be too expensive.

"The current (unlined) pits are necessary for efficient and economical operation of our drilling . . . operations," a Texaco manager wrote in the 1980 internal letter. "The total cost of eliminating the old pits and lining new pits would be $4,197,958. . . . It is recommended that the pits neither be . . . lined nor filled."

Also during the trial, a former minister of Ecuador's Ministry of Mines and Energy—the government division responsible for oil drilling—testified that a Texaco subsidiary knowingly used primitive waste disposal techniques in the 1970s and 1980s.

Texaco, called Chevron Texaco since its 2001 merger, always claimed it did nothing wrong and simply followed the standard environmental practices of the day. Furthermore, the company insists, most of the pollution occurred after it had handed over day-to-day operations to Petroecuador.

Technically Texaco is right; it may well have upheld Ecuador's environmental rules at the time. But Ecuador *didn't have* any environmental legislation back then. The Oriente was virgin rain forest and Ecuador's government depended on Texaco to apply the same environmental standards it used in the U.S. Without Texaco, Ecuador would not have been able to get its oil industry off the ground and so it deferred to the company. In the U.S., oil wastewater had been routinely re-injected into the ground for at least two decades but Texaco chose not to adopt this practice. By choosing to dump the water in unlined pits (a practice that was outlawed in the Texas oil fields as early as 1919), Texaco saved itself $3 per barrel, or roughly $4.5 billion over the life of its operations in the Oriente.

In 1995, under pressure from an Ecuadorian government that was itself feeling the heat from the damage done to one of the most pristine parts of the nation, Texaco agreed to finance a limited cleanup of 139 of the more than 600 toxic pits it had dug. Yet the $40 million effort made no provisions to remedy the hundreds of oil spills caused by the company or to monitor the health of the communities affected by Texaco's operations. And, claim opponents of Texaco, the company hired to conduct the cleanup often filled the toxic pools with dirt, forcing the wastewater further into the water supply and making pollution worse.

Texaco's drilling practices and the way it conducted community relations had a more profound effect than simply fouling up northern Ecuador—it established a modus operandi for other oil companies that continued to operate in the Amazon. Following Texaco's lead, company after company from Colombia to Peru laid waste to the rain forest while indigenous populations, many of whom had only minimal contact with the outside world before the oil companies showed up, have seen their land taken from them or have been cheated into giving up land in outrageous company-brokered scams that would make Peter Minuit and his beads-and-trinkets purchase of Manhattan blush.

Often oil companies will seek to divide indigenous communities by playing one village against another. Today, the companies employ social anthropologists whose job is to win over indigenous communities with promises of running water, electricity, and maybe the construction of a school or community health clinic. This allows the companies to claim they are actually improving the lives of the local people where they drill and even providing services that local governments don't provide. Sometimes the companies' efforts are genuine, but all too often these tokens come at a high price for the local communities. Many indigenous communities are spread out over a large area of the rain forest and have little chance to discuss the pros and cons of having oil companies drill on their land. Time and time again, once one section of a community has accepted help, the company makes them sign a contract allowing extensive drilling rights that represents a pittance in comparison to the real worth of the land and its resources. The company then claims that this contract represents the wishes of the whole community. It's a classic divide-and-conquer strategy and it continues to be applied throughout the Amazon basin.

As author Joe Kane vividly documented in *Savages,* his first-person account of the Huaorani indigenous nation's fight to stop U.S. oil companies drilling on their land in southeastern

Ecuador, the matchup between multinationals and rain forest peoples is a very unfair fight. On the one hand you have the companies, many of which have budgets and resources larger than those of the countries they are working in. On the other you have isolated tribes, most of whom have never seen a telephone, nevermind used one to call a good lawyer. Kane's book was published in 1995 and was a bestseller. Readers in the U.S. were shocked to see how U.S. oil companies like Maxus and Conoco were conducting themselves in such an environmentally sensitive part of the world and how they were doing things they would never have tried in the U.S. But *Savages* (no reader was left in any doubt who Kane was referring to in the title) also threw a very public light on an important new sensibility that had taken root in Ecuador's Amazon basin. After two decades of seeing their land decimated by oil companies, indigenous tribes ranging from the Quichua to the Shuar to the Huaorani had vowed to fight back.

Birth of an Indian Nation

An active indigenous movement had been flourishing in Ecuador since the early 1990s, but it was mainly confined to Ecuador's Quechua highland Indians who rarely had much contact with the lowland tribes of the Amazon. But by the late 1980s, a coalition of Amazon nations had joined a sometimes uneasy alliance with the highland Indians to form a united front against the oil companies and Ecuador's government. The Huaorani and other oil-threatened communities were also winning the support of a growing number of U.S. environmental organizations and the new grassroots movement of Ecuadorian environmental and human rights activists.

Together, these homegrown and international activist and political groups taught the Amazon communities about their rights under national and international law and how to understand the true value and implications of the deals they sign with oil companies. They also showed how to deal with companies on their own terms—by making use of national politics, the legal system, and most of all, the bully pulpit of global public opinion to ensure oil companies remain accountable for their global operations.

The Amazon Indian oil fights of the early 1990s struck a nerve with people around the world. Environmental and human rights issues, especially those associated with the Amazon, where there was a general consensus that "the lungs of the planet" were being destroyed, had taken on a fresh importance with the twentieth anniversary of Earth Day in 1990. One year before, the *Exxon Valdez* had run aground in Alaska's Prince William Sound, and by 1992, as celebrations began all through the Americas marking the five hundredth anniversary of Columbus's journey to the New World, indigenous groups throughout the continent were reenergized to fight for their individual rights.

Ecuador's Indian groups make up one-third of the country's population. While they couldn't prevent all new oil drilling in the Oriente, they made life very difficult for companies like Conoco, Maxus, and Burlington Resources, which had been granted concessions by Ecuador's government but faced volatile and often violent opposition from Indians in the rain forest.

In 1998, Ecuador's indigenous leadership registered its greatest success when it forced the Ecuador government to amend the national constitution. The new rule of law recognizes the existence of indigenous peoples as members of the "indivisible Ecuadorian state" and allows them to "define themselves as nationalities that have ancestral roots." Implicit in this new language was the Indians' right to protect their territory and natural resources.

In recent years, indigenous groups like the Shuar and Achuar have used this law to keep oil companies from drilling on their lands. In August 1999, some 400 Shuar and Achuar Indians marched on a provincial town in the south of Ecuador and delivered a legal injunction suing the Atlantic Richfield company (ARCO) in Ecuadorian court. The injunction called on the court to prevent ARCO from approaching individual Shuar communities or families with drilling agreements. It claimed ARCO was violating the newly amended Ecuadorian constitution by avoiding negotiating with the main elected Shuar and Achuar political authorities. They won and in late 1999 ARCO sold its Block 24 concession to another U.S. company, Burlington Resources, before pulling out of Ecuador altogether.

Nowadays, Ecuador's indigenous groups are in contact with other oil-threatened communities throughout Latin America, often with the organizing and logistical help of environmental groups like Amazon Watch, Earthrights International, and the Center for Economic and Social Rights. Veterans of Ecuador's oil wars have traveled to isolated communities like the Machiguenga in southern Peru, to teach them the lessons of how oil companies will treat their virgin rain forest land. The risks are great. U.S. and Argentinian oil companies believe the Camisea region, which the Machiguenga call home, has some of the greatest natural gas reserves on the planet. Today, environmentalists consider Camisea to be one of the most destructive oil and gas projects under development in the world.

At the same time, Ecuador's indigenous people are still fighting a rearguard battle against oil companies and their own government, which routinely ignores their territorial claims in order to reward lucrative drilling contracts to the oil companies.

By far the biggest threat comes from the Oleoducto De Crudos Pesados (OCP), a new pipeline that connects the Amazon fields to the Pacific coast. The pipeline is intended to help Ecuador double crude oil production to 800,000 barrels a day and provides a huge incentive for new oil exploration in parts of the rain forest that so far had been considered too environmentally sensitive or where the companies had previously shied away from a clash with indigenous groups that oppose them.

One of the main supporters of the pipeline and new drilling is the International Monetary Fund. Like many other oil-rich developing nations, Ecuador owes crippling debts to the international financial community and can only meet its debt payments by tapping more of its natural riches.

And here is where the blinkered approach of the international financial community becomes apparent. The IMF insists that the majority of new oil revenue that comes from the OCP be used to meet Ecuador's debt payments. That means the government faces great pressure to grant new oil concessions in the Oriente without any real hope that the oil money will be invested in improving the lives of Ecuador's people. Instead, Ecuador is locked into a vicious cycle. It has to keep drilling for oil simply to service the enormous debt it owes from relying too much on oil in the first place.

Shell Game

Throughout the 1990s the environmental movement in the U.S. and Europe began to pay closer attention to the business practices of oil companies around the world. The actions of companies like Texaco pointed to an industry that placed profit firmly ahead of environmental and human rights concerns, but it wasn't until 1995 that an oil company would be publicly humiliated all around the world. The company was Royal Dutch Shell and even hardened industry hands were appalled to see the company stand by as Nigeria's military dictatorship arrested then executed Shell's most prominent and vocal critic, Ken Saro-Wiwa.

Nigeria is the most prolific oil producer in Africa. Since

joining OPEC in 1971, it has consistently ranked among the world's top ten largest producers of oil and has earned billions in revenues from production agreements with oil companies such as Shell and Chevron. Nigeria's economy is heavily dependent on oil, which brings in some $20 billion each year and accounts for 90 percent of export revenues and nearly 80 percent of all government revenues. And how has that money helped Nigeria's population? Today, 66 percent of Nigerians live below the poverty line, and although Nigeria received over $300 billion in oil revenues over the past twenty-five years, per capita income is less than $1 a day. As the World Bank describes it, "Economic mismanagement, corruption, and excessive dependence on oil have been the main reasons for the poor economic performance and rising poverty."

Politically, Nigeria is a disaster. Since becoming independent from the United Kingdom in 1960 (three years after the first oil was tapped) the country has been under military rule for twenty-eight years. It has fought one major civil war—the Biafra conflict of 1967—spurred by ethnic tensions over its oil resources and for the past decade there has been bloody civil conflict between a number of ethnic groups in the main oil-producing regions of the Niger delta and the Nigerian military.

Meanwhile, billions of dollars have been spirited out of the country into the bank accounts of Nigeria's ruling elite, including approximately $2.2 billion absconded by Nigeria's former military dictator, Sani Abacha, during his five years in power from 1993 to 1998. According to Transparency International, an independent group that tracks corruption around the world, Nigeria ranked last but one out of ninety-one nations it surveyed in 2002. (Bangladesh had the dubious honor of coming ninety-first.)

Nigeria is one of the most egregious examples of an oil-driven economy out of control but it is hardly on its own. Just like Ecuador, Nigeria, along with all of the major oil-producing states that make up OPEC—Algeria, Iran, Iraq, Kuwait, Libya, Qatar, Saudi Arabia, and the United Arab Emirates—are all dependent on oil revenue for economic survival. Ecuador was part of OPEC for eighteen years before it withdrew in 1990, and, though politically unstable, it can at least boast a fragile democratic tradition even if the military always lurks just one potential coup away. The remaining members of OPEC, however, along with most other oil-dependent nations, have no democratic government and a dismal human rights record. Consider non-OPEC nations such as Angola, Azerbaijan, and Kazakhstan where the potential of billion-dollar oil contracts has made state corruption a cottage industry. Or Sudan, where the discovery of oil added a new level of ferocity to an already barbaric civil war.

Africa's malaise is likely to be compounded over the next decade by full-scale oil exploration in the Gulf of Guinea, often referred to as the African Persian Gulf. International companies are flooding into this region and, soon, oil will bring more revenue to Africa than the continent has ever seen before—ten times the amount that Western donors give each year in aid.

Vast oil revenues allow governments to do away with levying taxes and that makes for very lazy democracy. The governments of these nations have no need for domestic taxation and so don't consider themselves accountable to their own population. In the short term, the citizens of the country are happy to avoid paying taxes. But over time, the general population becomes disenchanted with the government, especially when it realizes a tiny elite is creaming off all the cash.

That's where the repression effect, as political scientist Michael Ross calls it, comes into play. Oil-rich states spend a great deal of their natural-resource income on strengthening the military to counter threats from abroad and on domestic security to beat down challenges from within. One World Bank study showed that oil-rich nations were 40 percent more likely to fall into civil war than nations that didn't have significant oil resources.

Oil-driven corruption and repression occur on a daily basis all over the world. Most of this abuse goes unnoticed by the mainly Western consumers who reap the benefits of foreign oil. That's why the revelations about how the Nigerian military killed its own people in order to protect Shell's operation in the Niger Delta were so shocking to people all over the world.

Ever since the first Nigerian oil was discovered in 1957, Shell had been at the forefront of exploration in the country. By the early 1990s it was the leading multinational oil company in Nigeria and it had developed a mutually beneficial working arrangement with a succession of military dictatorships, all of whom had become experts at gorging themselves on the nation's oil profits. Meanwhile the people living in the main oil-producing regions on the Niger Delta, one of the world's largest wetlands, struggled in environmental squalor and poverty. One area of Shell's operations was Ogoniland, a small strip of land that was home to the Ogoni people, who made up a mere half a percent of the total Nigerian population.

Ogoniland contributed only 3 percent of Shell's Nigerian production but it had produced some $30 billion worth of oil since operations began there in 1958. The decades of drilling had decimated the region. Residents were forced to farm among oil spills and the constant fumes of gas flares. One environmental study found drinking water in one Ogoni village where Shell operated had petroleum hydrocarbon levels 360 times higher than levels allowed by the European Community.

In 1990, the Ogoni formed the Movement for the Survival of the Ogoni People (MOSOP) to fight for the protection of their land. They presented the Nigerian government with the Ogoni Bill of Rights, calling for political representation and the right to protect the land on which they lived. MOSOP's spokesman was Ken Saro-Wiwa, a leading Nigerian novelist and cultural figure. "Shell has waged an ecological war in Ogoni," he declared. "Human life, flora, fauna, the air, fall at its feet and finally the land dies."

On January 4, 1993, Saro-Wiwa addressed a protesting crowd of some 300,000, three-fifths of the total Ogoni population, and declared Shell to be "persona non grata" in Ogoniland. More disturbing for the company, MOSOP succeeded in shutting down Shell operations using nonviolent demonstrations throughout the region.

The Nigerian government organized a new security force, the Rivers State Internal Security Task Force, to counter MOSOP and was ruthless in suppressing the protesters, beating thousands of local people and executing hundreds of suspected MOSOP members. Shell had long relied on the Nigerian security forces to protect its operations from Ogoni protests. In 1990, a Shell manager made a written request for protection and the security forces responded by killing eighty unarmed Ogoni citizens.

Then in 1994, a Rivers State Internal Security Task Force major wrote a memo that suggested the military was prepared to crush resistance to the company. "Shell operations still impossible unless ruthless military operations are undertaken for smooth economic activities to commence," the officer wrote. Ten days later, Ogoniland was under military occupation and Ken Saro-Wiwa and eight other MOSOP members were arrested. They were accused of murdering four Ogoni chiefs—charges that all outside observers considered trumped up by the Nigerian government. Shell flatly denied having anything to do with the actions that led to Saro-Wiwa's arrest.

For eighteen months, Shell sat back as Saro-Wiwa and his colleagues were held in prison. They were then found guilty of murder by a military court and sentenced to death. Only after coming under withering criticism from human rights groups and the media did Shell call on Nigerian leader Abacha to commute the sentence to life imprisonment. But on November 10, 1995, Saro-Wiwa and his eight compatriots were executed.

Subsequent revelations that Shell had maintained its own police force and had tried to purchase $1.2 million worth of arms for this force only intensified criticisms both inside and outside the company that Shell's Nigerian arm had acted above the law. As one Shell executive who was not involved in Nigeria later put it, "The mistake we made is that we became the government."

The execution of Ken Saro-Wiwa shook Shell to its corporate core. The London-based parent company delegates a great deal of responsibility to its individual operating companies in different countries and there was little doubt that the Nigerian subsidiary had let events spin very badly out of control. Yet at the same time, it was apparent that Shell's head office had failed to appreciate the gravity of the Saro-Wiwa case. Coming on the heels of Shell's attempt to scuttle one of its oil rigs, the heavily contaminated 4,000-tonne Brent Spar, in the North Sea, Ken Saro-Wiwa's death made Shell appear both callous and reckless in the eyes of many of its shareholders.

In 1998, Shell published a landmark and soul-searching report *Profits and Principles—Does There Have to Be a Choice?* The report outlined how the company intended to integrate social responsibility into its overall business strategy and it pledged "to ensure that our businesses are run in a way that is ethically acceptable to the rest of the world and in line with our own values."

How successful Shell's new approach has been is a matter of debate. There is no doubt that the Ken Saro-Wiwa incident seriously damaged the company's credibility, especially as Shell had always considered itself one of the more progressive oil companies. Today, the company spends as much as $60 million a year on community projects in the Niger Delta and, in general, it has embraced a new high-profile commitment to socially responsible business. Nevertheless, a 2001 internal Shell report written by independent consultants found the company was still making mistakes in Nigeria. According to the report, the community projects lacked any real commitment to local people. "[Shell] development activity remains 'top down' and consultation extends only as far as the local elite," read the report.

Shell's pledge to honor human rights and promote sustainable development has not been embraced by the other major oil companies operating in the delta region. In 1998, Chevron (now ChevronTexaco) flew the Nigerian navy and the mobile police to one of its offshore drilling platforms that had been occupied by 200 protesting youths. Two unarmed protestors were shot dead when the military moved in. The following year, according to a 1999 lawsuit filed in California court, Chevron again aided a military operation to squash community protests against the company that also resulted in the deaths of civilians. Meanwhile, other operators, such as the French company ELF and Italian AGIP's Nigerian subsidiary, have also been accused of funding groups to beat and physically intimidate local anti-oil activists.

When local people see their land being ruined by oil spills, lucrative jobs being taken by foreign workers, and the much-promised rewards of "black gold" vanishing into the ether, they obviously blame their own government. But they also harbor a special contempt for the foreign companies that are drilling the oil.

That is unfair, say the oil companies operating in the delta. Companies like Chevron, Exxon, and Shell argue that they are bearing the brunt of community criticism that is really directed at the Nigerian government—over the years, the government has taken billions of dollars from the delta operations without investing back into those communities where oil exploration has had the most impact. In the past, the Nigerian government has also reneged on signed deals with the oil companies and has sought to play one company off another.

The companies also point out that they must operate in a region that has been ripped apart by ethnic fighting (most of it fueled by the inequalities of oil-wealth distribution) and that the companies and their employees are often targeted as pawns in the fight against the government. Offshore oil rigs are routinely stormed by anti-oil groups, which often hold oil workers hostage in order to get their message across. So it is understandable that the companies want to protect their workers and assets. But it is the oil companies' own choice to operate in countries like Nigeria. And when they align themselves so closely with the police and military apparatus of those regimes, the companies are ignoring the codes of conduct and ethics most Western citizens expect companies to operate under.

As a 1999 Human Rights Watch report on oil operations in Nigeria surmised, it is not enough for the oil companies to distance themselves from regimes they do business with. "The dominant position of oil companies in Nigeria brings with it a special responsibility in this regard to monitor and promote respect for human rights," the report read. "Given the overwhelming role of oil in the Nigerian national economy, the policies and practices of the oil companies are important factors in the decision making of the Nigerian government. Because the oil companies are operating joint ventures with the government they have constant opportunities to influence government policy, including with respect to the provision of security for the oil facilities and other issues in the oil-producing regions. All the oil companies operating in Nigeria share this responsibility to promote respect for human rights."

If oil companies' cynical approach to human rights and the environment was restricted simply to Nigeria, it might be easier to attribute their actions to the problems their local managers face operating in one of the world's most dangerous and corrupt countries. But this pattern of abuse and complicity with repressive military forces continues to occur all over the world.

Hit Them Where It Hurts

More than thirty years on from Texaco's disastrous entry into the Oriente, more oil is being drilled in environmentally sensitive areas than ever before. And while a few of the largest oil companies have made progress in how they deal with sensitive indigenous and local communities and how they manage the industrial footprint they leave on ecologically fragile parts of the world, the majority of oil companies still act as if it was business as usual.

Today, environmental activists tend to look at oil companies as three main groups. There are the northern Europeans— Shell and BP—which have taken some concrete measures to improve their commitment to human rights, the environment, and the pursuit of alternative energy.

Then there are the American majors, ExxonMobil and ChevronTexaco, along with a few smaller U.S. independent concerns, which talk a good game about environmental commitments but make no effort to change their practices. Finally, there are the oil companies from the rest of the world—a motley crew of independents and state operations whose concern for environmental and human rights issues is nonexistent.

The rush to drill will only accelerate as global demand increases and as more nations join the ranks of the industrialized world. And while the loose global network of environmental and human rights organizations has only a fraction of the political muscle of the major oil companies, it continues to sound a loud clarion call against oil-company excesses among a general public that has little sympathy for the methods of Big Oil.

Nothing has helped generate more support around the world for oil issues than the Internet. Using web sites and listserves as both a fundraising and mobilizing tool, U.S. and European activists have succeeded in building a loosely connected but highly motivated network of grassroots support.

There is no better example of this than the global campaign that came together in the late 1990s to help the U'wa people in northern Colombia whose lands were part of an oil concession granted by the Colombian government to the California company Occidental Petroleum.

Like many indigenous people, the U'wa believed any oil drilling would destroy the land and the rivers which they held sacred and worshipped. Rather than see their land poisoned, the entire tribe of 5,000 people made a pact. They decided that, if Occidental began drilling, they would climb the highest cliff on their land and commit collective suicide by jumping off.

Talk about a potential public relations disaster for Occidental. Yet the U'wa would have had little leverage with the oil companies or the Colombian government if they hadn't been able to publicize their predicament. After all, the U'wa have little contact with the outside world. They were, however, in contact with other indigenous nations in South America, and soon news of their plight reached environmental and human rights groups in the U.S. Before long, groups like Greenpeace, Rainforest Action Network, Amazon Watch, and Human Rights Watch were trumpeting the U'wa's problems and their vow to resist Occidental to a global audience in a flurry of e-mails, newsletters, and listserves on the Internet.

Occidental ultimately backed off from its original drilling designs, though the U'wa are still fighting any oil operations on their land. Now seen as a cause célèbre, members of the U'wa, funded by international nongovernmental organizations, started to take their message to other oil-affected communities and to politicians and opinion makers in the U.S. and Europe.

An important weapon in the fight against Occidental was the pressure activists placed on the company's biggest investor, Fidelity. After dozens of protests and unflattering press about the investment house's link to the U'wa, Fidelity dropped half their shares in Occidental.

Increasingly, activist groups are realizing that the most effective way to target large corporations is to hit their bottom line through consumer boycotts and shareholder protests. The reasoning is simple—the companies may not care about Amazon Indians, but they pay very close attention to the desires of their investors and shareholders.

In recent years, major international companies like Home Depot and Staples have agreed to embrace more environmentally conscious business practices after suffering PR nightmares at the hands of environmental groups and, slowly, corporate responsibility is also creeping into the oil industry.

One of the most successful activist campaigns is Publish What You Pay (PWYP), an appeal by some 130 nongovernmental organizations led by Global Witness for oil, gas, and mining companies to publicly disclose all rents, fees, royalties, loans, taxes, and any other payments to governments in return for extracting resources in those countries.

The initiative would target all nations whose governments receive large payments from oil and gas companies and if adopted would add a new level of corporate transparency. Consider Angola's state oil revenues for a moment. In 2000, some $1 billion of the $6.9 billion the country received in oil exports flowed back out of Angola and into private bank accounts. This despite a national law that stipulates the country's oil resources belong to all Angola's citizens. That year, the Marathon Oil Company of Houston, Texas, sent over $13 million to an offshore bank account owned by Sonangol, Angola's state oil company. The amount was part of a bonus Marathon had agreed to pay the Angolan government for drilling rights in the country. Throughout 2000, the offshore bank was used to transfer cash to, among other dubious places, a charitable foundation run by Angola's president.

"By not publishing what they pay," wrote Global Witness, "oil companies endorse a double standard of behavior that would be unacceptable in the North and make it impossible for ordinary Angolan citizens to call their government to account over the management of revenues."

BP was the first oil company to back the Publish What You Pay initiative. This so shocked the Angolan government that it threatened to cancel BP's concession before later relenting. The U.S. companies, ExxonMobil, ConocoPhillips, and Chevron-Texaco, along with French giant TotalFinaElf have so far balked at adopting the PWYP, claiming it breaches confidentiality agreements signed with the Angolan government and arguing that even if they disclosed what they paid, the opaque nature of Angolan business would still make accountability difficult.

Of course, this reluctance to anger the Angolan government might also have something to do with the huge importance that the companies, and the Bush administration, are placing on securing natural resources from this part of Africa.

Yet PWYP is gaining momentum. Nigeria and São Tomé have both agreed to disclose their new oil deals and Global Witness has won the support of the World Bank and International Monetary Fund. The British government has also thrown its support behind the plan. The members of the G-8 leading industrial nations have called for the adoption of the PWYP principles only on a voluntary basis, however. Global Witness points out that this is unlikely to occur when the autocratic leaders of so many resource-rich nations treat oil revenues as their own private treasure chest. It, along with financier George Soros and groups of other international organizations, has called on the major stock markets to establish new rules requiring all listed companies to disclose what they pay for corporate contracts as a term of remaining on the exchanges.

4

Boom and Bust: The Price of Oil and How Much Oil Remains

Behind the thick glass windows of the New York Mercantile Exchange (NYMEX) viewing gallery, the muffled cacophony of the trading floor seems quite comical. In the circular section of the trading floor devoted to crude oil—the pit, as it is called—the traders, nearly all of them men, gesticulate wildly, scream at each other and push each other out of the way as they compete to buy and sell crude-oil future contracts. To the uninitiated, it's like being at the zoo and watching some strange new species of animal indulge in its daily ritual.

The traders are dressed in brightly colored waist-length jackets with the name of their companies emblazoned on the back—the green of Rafferty Associates and the bright yellow of Goldman Sachs, the red of Top Energy, Inc., and the orange and black of BNP Paribas. The colors are important—amid all the lunacy going on down there, the traders want to make sure they don't try to cut deals with their own guys. Lea Meierfeld, a young member of the NYMEX public relations staff, slides open one of the huge panes and this strange tableau roars into life. What I'm hearing is the serious business of making money—quickly—and the sound is deafening: it is so loud that I'm amazed any work gets done.

Once they set foot on the trading floor, the traders enter a world running at hyperspeed with its own language and culture. The traders rely on a team of runners, known as clerks, to carry information back and forth to them from other brokers who sit on the fringes of the action. Once a trade is made, the brokers record it on a paper card and throw that card with a flourish into the center of the trading ring. There, four men sit and record each trade as it lands at their feet. They wear protective goggles to shield their eyes from the cards that fly in their direction. If the card isn't thrown cleanly into the recording area of the ring, the trade doesn't count and millions of dollars can be lost, so traders spend hours practicing the perfect throw and flick of the wrist, like fly fishermen perfecting their cast-off.

"The guys take classes to make sure they hit with the card," Lea tells me as we watch the bedlam from above. She should know; her whole family trades and has been since the days when the main business of the Mercantile Exchange was eggs and butter. "My brother and father used to practice throwing cards from the porch," she says.

For those of us not connected to the financial markets, one of the toughest concepts to grasp is that oil is governed by one global price. Essentially, it's a bit like pouring all the oil in the world into one big pool and then pricing it according to how full the pool is at any one time. If the pool is a little emptier than people expect, then the next barrel added can command a higher price than the last and send oil prices rising. If oil traders believe there is a surplus of oil flowing into the pool to meet demand, then the next barrel will not command the same value and prices will drop.

And because of the global nature of the system, it doesn't really matter where the U.S., for example, physically gets its oil. Even if all U.S. companies with agreements to supply U.S. consumers have contracts with nations like Canada, Mexico, or Russia, the price of that oil is still governed by market conditions elsewhere in the world. In theory, the U.S. might not get any oil from a country like Venezuela, yet a major disruption in Venezuela's oil supply from a national strike, hurricane, or coup d'état would still affect the global supply of oil and drive up the price for everyone.

Traditionally, all international oil contracts were conducted at a fixed price between the so-called upstream oil producers and the downstream refiners and retailers. That's because, for the first century of oil, the major companies had a stranglehold on production. A company like BP would produce oil in Iran and then sell it either to one of its own refineries or to another company or directly to a Western government. The Iranian government got a percentage of the sale but it didn't get to set the price. It was in BP's interest to keep that price stable through controlled production quotas—a lot lower than Iran might like—because none of the major companies wanted to spark a production surge that could drive down prices. A stable price also helped the main oil-consuming nations regulate and plan their economies.

But ever since the Middle East producers threw that cozy old system out of whack by establishing OPEC, buyers have had to factor in the element of surprise. Almost overnight, following the first OPEC price hike, a new type of oil transaction—the spot market—sprang up. The spot marked presented oil for sale based on the rate of the day. It offered greater transparency (if not a great price) to the companies that were getting burned by being locked into fixed contracts.

Paying spot market prices tends to be the most expensive way of buying oil and, because of this, only a small amount of oil deals are conducted on the spot market. Most oil is still secured and delivered under terms of fixed contracts between producers and refiners. But because oil supply is so susceptible to world events, be they work stoppages, war, natural disasters, or political instability, both buyers and sellers need some protection against being locked in on those contracts. That's where the oil futures exchange comes in and that's why the NYMEX, the most actively traded crude oil futures market in the world, is the global bellwether for setting the price of oil.

A futures contract is a promise to deliver a given quantity of a standardized commodity at a specified place, price, and time in the future. No oil physically changes hands but daily and hourly patterns of supply and demand help establish a realistic market price for oil. The NYMEX records all transaction prices then posts them on the Internet. This price transparency allows companies to negotiate a realistic price for oil contracts based on current market conditions. It also lets them "hedge" against price fluctuations at a later date by locking in the current market price for future crude deliveries.

The NYMEX is one of three major oil futures exchanges, the others being the International Petroleum Exchange (IPE) in London and the Singapore Exchange. Each exchange selects one type of crude oil to represent the base price for its regional market (the three represent the major refining centers of the U.S. Gulf Coast, northwest Europe, and Singapore) and then adds or subtracts a quality adjustment in relation to the base crude for every other type of crude being traded. The NYMEX, for example, uses the price of West Texas Intermediate as its base crude because it is the most common and high-quality form of crude to flow through the important Texas Gulf Coast ports—where most oil enters the U.S. In the same way, crude oil sold on the IPE is tied to the price for North Sea's Brent crude, while crude sold on the Singapore exchange is tied to Dubai crude from the Middle East. The price adjustment is determined by the quality of the crude and how easily it can be refined into high-value products.

At the NYMEX, thousands of crude oil transactions are conducted daily but few of these shipments are ever delivered. Instead, they are constantly retraded based on the current market price. In this way the futures market works as a barometer of the right price of oil contracts in the future, based on all the information available now.

And because of that, every NYMEX trader you will ever

HITTING BLACK GOLD:
HOW COMPANIES DRILL FOR OIL

Crude oil formed hundreds of millions of years ago when dead plants and tiny plankton-like animals settled on the bottom of lake beds during the Jurassic period. Stagnant water preserved the debris from oxidizing, and the hydrogen and carbon matter was converted by heat and pressure into hydrocarbons—oil and gas—below the earth's surface.

Computer technology and seismic mapping have revolutionized the search for oil. Using huge thumper trucks to cause vibrations on land, and heavy-duty air guns to bounce a seismic signal off ocean floors, companies build a digital 3-D map of potential oil fields. Once oil is located, companies use rotary drills to penetrate the ground, typically to depths of around one to two miles.

When oil first flows from a field, it gushes out under great pressure. But, as oil fields mature, there is less underground pressure to force oil to the surface. It's like a balloon losing air except that the porous rock in the reservoir doesn't shrink. Instead, the oil expands to fill the rock and so doesn't flow to the surface as quickly as before.

Oil companies use a variety of techniques to breathe new life into mature fields. They limit the number of wells they drill in order to maintain maximum pressure underground and they inject highly pressurized fluids into the reservoir, creating a crack in the rock that helps the oil flow. Companies also inject carbon dioxide gas to reduce the oil's density and help its flow in a process known as carbon sequestration.

Perhaps the oil industry's greatest technological success so far is horizontal drilling. Drilling down first vertically and then horizontally allows companies to tap reservoirs where the oil lies in only a thin strip, like the icing between two layers of cake. Drilling a vertical well into such a reservoir would access just a small part of the field. A horizontal drill can reach oil lying 1,000 feet or more on either side of the vertical drill.

meet is a news junkie. On the floor of the exchange, along-side the monitors showing the constantly changing price of new oil futures, sit TV sets tuned to CNN, Fox News, and CNBC. Be it war in the Gulf, a general strike in Venezuela, or ethnic conflict in Nigeria, major news events affect the trader's confidence. Most important for them, the day's news determines whether they make or lose money.

Why the Middle East Matters

Retail gas prices have an important economic and psychological influence on the spending patterns of the U.S. consumer. Europeans pay far more for gas than American consumers due to the higher rates of taxation their governments demand. But as dependent on oil as Europe is, the U.S. is in a far more precarious position. Americans own more cars than the rest of the world and depend on them to travel long distances for all aspects of their daily life; so a sharp rise in the price of crude oil hits the economy across the board.

America's 191 million drivers see the posted price of gas every day. Americans discuss the ever-changing price of gas as often as they talk about the weather. Just like a month of rainy days can get you down, a month of rising gas prices can have a disproportionate negative effect on consumer confidence.

Nearly half the cost of a gallon of gasoline is directly attributed to the cost of crude oil. So not surprisingly, the U.S. has worked hard over the past twenty years to minimize its reliance on the largest and most prolific regional producer of crude oil, the Middle East. But even while the U.S. government touts new domestic oil drilling in Alaska and the western Rockies, and while the major oil companies broker new oil partnerships with Russian oil companies as well as feverishly increase deepwater exploration, both the government and the companies understand that the fate of the oil business will continue to be governed by events in the Middle East.

The Role of OPEC

When OPEC first raised oil prices and then nationalized Western oil company assets, oil prices more than doubled and the world economy fell into a prolonged recession. Even when the developed world bounced back, most developing nations still suffered—unable to meet the high prices, they had taken out huge loans from foreign banks to keep their oil-dependent economies moving only to find themselves paralyzed by the interest payments on those loans.

Nationalization was an immense blow to the major oil companies—all of whom had invested nearly half a century in the Middle East (while making a killing in the process, of course). Now they had to search for alternative sources of oil and their focus turned to exploration in Asia, Africa, northern Europe, and Latin America. But, thankfully for U.S. consumers, OPEC quickly squandered its stranglehold on the world oil market with greedy overproduction. By the mid-1980s the price crashed to under $10 a barrel.

Today OPEC nations account for just about 30 percent of

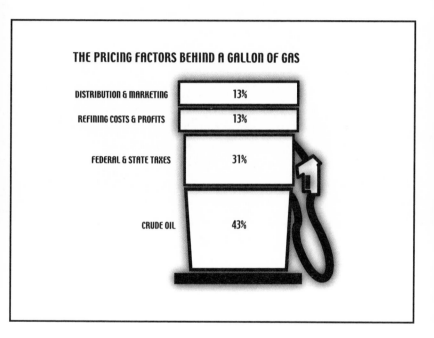

world oil supply and about 40 percent of total U.S. imports, down from 70 percent in 1977. Thirty percent isn't enough market share to control the price of oil and so OPEC oil must compete against oil from the rest of the world—especially, in recent years, cheap Russian oil—and that means no one nation or company can unilaterally control prices.

But that 30 percent still gives OPEC a strong say in the price of oil. OPEC members, most of whose national budgets are almost completely dependent on oil revenues, cannot afford to see oil prices fall too low. Conversely, OPEC fears that high prices will prompt U.S. and other international companies to start new oil exploration thus further diluting its market share. High oil prices could even force the major industrialized nations to get serious about developing alternative energy.

As Saudi oil minister Sheik Yamani said in 1981, "If we force Western countries to invest heavily in finding alternative sources of energy, they will. This will take them no more than seven to ten years and will result in their reduced dependence on oil as a source of energy to a point which will jeopardize Saudi Arabia's interests."

For that reason, OPEC has long looked to maintain an ideal market price of $25 a barrel for its oil though it is content to see prices fluctuate between $22 and $28 a barrel. If market prices rise above or below its target range for any extended period of time, OPEC can increase or cut back production to bring prices back in line.

Officially, OPEC has a mechanism that regulates its basket of prices by automatically triggering a cut of 500,000 barrels per day when the price drops under $22 for ten consecutive days and an increase of 500,000 barrels per day when the price stays above $28 for twenty consecutive days. In practice, however, the regularly scheduled meetings of OPEC energy ministers in Vienna really decide the cartel's price and production strategy.

OPEC's price basket might keep world oil prices higher than the real open market would allow but it has also reinstated a level of price stability and sanity to a global economy that doesn't need any more oil shocks. Recently, though, OPEC ministers have suggested they may be ready to abandon their quotas and let prices rise in order to recoup the losses they've experienced due to the low exchange rate of the U.S. dollar.

One OPEC member, Saudi Arabia, plays a special role. Only Saudi Arabia has the sufficient spare capacity (some two million barrels a day) to counter a production shortfall elsewhere in the world. If a major oil-producing nation has its production cut by a war, strike, or natural disaster, Saudi Arabia plays the role of honest broker and increases its daily production to make up for the loss. Why does Saudi Arabia do this? Part of the reason is its close relationship with the U.S. Another part is that Saudi Arabia has long maintained a very conservative attitude to oil production. It warned against OPEC overproduction in the early 1980s and it continues to advocate a cautious approach to keeping prices steady. It may have the greatest reserves of oil in the world but Saudi Arabia is as dependent on oil as the U.S. While long-term high prices cripple the U.S., long-term low prices do the same to the Saudi kingdom.

The Future of Oil

So, are we running out of oil?

It's a concern that has dogged the oil industry since its early days when the seemingly never-ending riches of a gushing new field could dry up overnight. Improved geological techniques have taken a lot of the guessing out of oil exploration but the boom-or-bust mentality persists in the industry to this day—there is still an urge to pump new fields quickly, to make money before the well runs dry.

Most experts agree that we are not running out of oil—there is enough of the black stuff lying below the earth's surface to last well through the end of the twenty-first century. But worrying about how much oil is left misses the point. It is the price of oil, not how much physically remains, that will determine the future course of U.S. foreign, economic, and energy policy.

In 1956, a geologist for Shell called M. King Hubbert published a paper on oil production in the lower forty-eight states (Alaskan oil had yet to be commercially explored). Hubbert used a simple bell curve to illustrate his findings. He argued that oil production grew slowly then accelerated sharply as oil companies discovered major oil fields. But after the biggest fields had been tapped and exploited, overall production slowed. The point at which production began slowing came at the top of his bell curve and was what he called peak production. From that point on the production curve fell

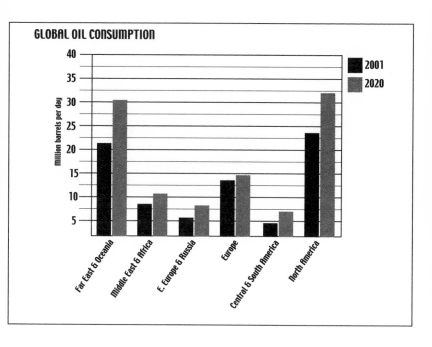

GLOBAL OIL CONSUMPTION

Million barrels per day

■ 2001
■ 2020

Far East & Oceania
Middle East & Africa
E. Europe & Russia
Europe
Central & South America
North America

as sharply as it rose as only smaller new fields were discovered and they became increasingly more expensive to exploit.

Based on this research, Hubbert forecast that oil production in the lower forty-eight states would peak between 1965 and 1970 even though the U.S. was then producing more oil than it ever had before. It turns out he was right. In 1970, crude oil production in the lower forty-eight states reached 9.4 million barrels a day. Aside from Alaska, where production peaked in 1988, domestic production levels have been steadily declining ever since.

The Energy Information Administration (EIA), a research arm of the U.S. Department of Energy, projects that world oil demand will increase from 78 million barrels a day in 2002 to 118 million barrels a day in 2025. Some of that demand will come from the U.S. where net imports will account for 70 percent of total U.S. petroleum demand by that time. But most of the growth will come from the rapidly expanding oil-thirsty nations of the developing world, notably China. In 2003, China accounted for 35 percent of the total growth in world oil demand. That pattern could continue for some time.

Now, as long as new discoveries of oil around the world continue to match the rise in demand, there isn't really too much of a problem, at least price-wise (the effect of all that oil on global warming is a whole other thing). Even if global demand outpaces new discoveries in the short term, countries like Saudi Arabia, Iran, and Iraq still sit on large amounts of untapped reserves.

But when world oil reaches the point of peak production and demand starts to outpace all possible supply, then the price of oil will start a slow but continuous rise and the global economy will start to feel the strain. That would be disastrous for all industrial nations, but especially for the U.S.

A Collision Course

No one inside or outside of the oil industry denies that world oil output will someday reach peak production. The important question, of course, is when will that happen?

In 2000, the United States Geological Survey (USGS), having conducted a report on the world's known oil reserves, estimated that 3,003 trillion barrels of recoverable oil remain in the earth. Using the USGS figures, the EIA estimates that global peak production is still some thirty-three years away.

Industry boosters like the American Petroleum Institute, along with the major companies, maintain that ample new streams of oil remain to be discovered in the world, especially in deep ocean deposits. And they say that new technology like 4-D exploration and horizontal drilling will help oil companies obtain more oil from existing reserves than was ever thought possible. They also point out that exploration and production of heavy or nonconventional oils—like bitumen-based shale and tar sand—will become increasingly cost competitive in the future (especially as the price of conventional oil rises), adding potentially billions more barrels of recoverable oil to global reserves. The combination of new technology and new markets for heavy oils, they say, could delay peak production for many years.

But a small group of industry geologists offers a vigorous challenge to these predictions. In 2003, the Association for the Study of Peak Oil and Gas (ASPO), a collection of Hubbert acolytes, issued a report saying that peak oil production could occur as early as 2010. ASPO's founder, former oil company geologist Colin Campbell, believes the USGS figures could have overstated global reserves by as much as one-third.

Skeptics like Campbell contend that oil-producing nations, especially those in OPEC, routinely inflate their own oil reserve estimates in order to obtain International Monetary Fund or World Bank loans secured against their future production levels. In the mid-1980s global proven crude reserves were said to be between 650 billion and 700 billion barrels. Ten years later, an extra 300 billion barrels had been added to the global reserves even though no new major oil fields had been discovered. Saudi Arabia, for instance, managed to add close to 100 billion barrels to its stated reserves in 1990 alone.

Different oil-producing nations also apply very different standards when it comes to judging how much oil is ultimately recoverable. Indeed, more than half the nations included in *Oil and Gas Journal*'s annual report on oil reserves—the industry standard source for calculating oil reserves—routinely claimed to have the same amount of proven reserves as the previous year even though they were tapping three times more oil on average than they were discovering.

Then there is the question of what is recoverable oil. Take Canada. Depending whom you talk to—and the estimates vary quite considerably—Canada is sitting on a mother lode of oil. In 2003, the U.S. government elevated Canada to second place (after Saudi Arabia) in its rankings of nations with the greatest oil reserves. According to *Oil and Gas Journal,* Canada has recoverable reserves of 180 billion barrels even though the vast majority exists as tar sand in Alberta province. Alberta has abundant quantities of tar sand and about 9 percent of the deposits lie close to the surface, allowing a recovery rate of probably 50 percent perhaps stretching up to 90 percent. Oil companies are salivating that at least some of the tar sand—or oil sand as the companies prefer to term it now (maybe between 174 billion and 270 billion barrels)—will be recoverable to be refined into gasoline and other oil products.

In Venezuela the national oil company believes some 1.2 trillion barrels of nonconventional oil lie just a few thousand feet below the surface of the Orinoco River's north shore. At present only 5 percent of this would be considered recoverable but in a few years new technical improvements could raise those estimates to 25 percent—a possible 300 million barrels of refinable crude oil. The United Nations Institute for Training and Research Centre for Heavy Crude suggests that the amount of recoverable heavy oil in the world might equal one-third of the total known conventional global oil reserves. Aside from those in Canada and Venezuela, large tar sand deposits have also been identified in Australia, Brazil, China, Estonia, and the U.S.

Total heavy-oil production only adds up to about 3.5 percent of total world oil production but the technology needed to effectively separate the petroleum from the sand is improving all the time. One company, Genoil, has developed a process that upgrades heavy oil to light oil by adding hydrogen and increasing the yield of light oil by 30 percent over conventional techniques. Syncrude Canada, which operates the largest tar oil operation in the world, is meeting 13 percent of Canada's total oil needs and by the end of the decade Canada and Venezuela hope to produce one million barrels of oil per day.

Up until now, there has been very little incentive to access heavy oil because the process is so laborious and expensive. It takes two tons of tar sand to make one barrel of crude oil. Vast areas of land have to be mined from open pits and then the rocks crushed to reach the point where oil can be extracted. These mini-particles are then treated with hot water or solvents to release the oil from the sand and then the oil must be further refined into fuel oil. All told, producing tar sand oil costs Syncrude Canada about $12 a barrel. That is competitive with the most mature of America's depleted oil fields but is still no match for Saudi Arabia's $1-a-barrel production costs.

Industry analysts says this type of tar sand operation is commercially viable if oil prices stay above $18 a barrel, but what really becomes attractive is if the cost of oil goes up to $35 or $40 a barrel and remains there. The United Nations report on heavy oil suggests this scenario and says, "the peaking of conventional oil, combined with physical shortages and crude price escalation [sometime in the second decade of the coming century] will mark the beginning of commercial production from the world's huge deposits of tar sands and extra heavy oil."

Few nations have previously included nonconventional oil in their reserve calculations specifically because it is so expensive to produce. But, despite the economic and technological uncertainty surrounding unconventional oil production, the EIA forecast for future global oil consumption assumes that increased nonconventional oil production, predominantly from Canada, will counter a decline in conventional oil production in the United States, Canada, Mexico, Western Europe, and Australia. Using this accounting, it is possible to argue that there will always be plentiful supplies of oil in the world. But if our best hope of meeting new global demand lies in tapping nonconventional oil, then the age of cheap oil surely is over.

Since the 1970s, the U.S. has been successful in cutting its dependency on Middle East oil and it has targeted a number of nations to help protect its energy security. In recent years, the U.S. has been happy to see Russia ramp up production—so much so that in 2002 it was producing seven million barrels each day, briefly displacing Saudia Arabia as the world's leading producer. President Vladimir Putin's government depends heavily on energy revenues and so is eager to pump all the oil it can. Russian oil companies have invested heavily in updating their old infrastructure and the big Western companies have been keen to strike partnerships with them. In June

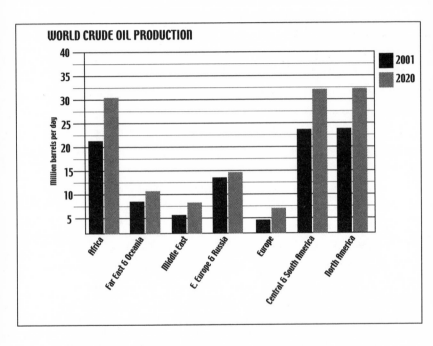

2003, BP announced a partnership with the Russian oil giant TNK and both Exxon and ChevronTexaco have wooed Russian companies in the hope of tapping into their oil reserves. At the same time the Russian government has encouraged a flood in oil production.

Both the U.S. and Europe have embraced Russia's new role because they see Russia acting as a strategic counterweight to the Middle East states. But how dependable are Russian oil supplies? Pessimists contend that the U.S. government is overestimating Russia's potential. In 1975, the Russian fields reportedly held 83 billion barrels of oil. Today the EIA says Russia holds reserves of 60 billion barrels. Part of that discrepancy can be attributed to different accounting methods or data mix-ups that followed the breakup of the Soviet Union, but the overall picture, say the skeptics, is clear. Russian oil might not have the staying power the U.S. is hoping for.

Neither can the West depend on the major oil fields that helped it shed its Middle East ties after the 1970s. North Sea oil peaked at the end of 2002 and the major Prudhoe Bay field of Alaska peaked back in 1988. Even the Caspian basin now seems destined to be a contender rather than a heavyweight champ. According to estimates by the Oil Depletion Analysis Center (ODAC), a London-based research group, the Caspian Sea contains some 50 billion barrels of ultimately recoverable oil—a large amount but only the size of the North Sea reserves. At projected rates of production, it could peak in production by 2010.

Despite the tremendous amount of energy that has been

spent scouring the world for new oil, geologists now admit that most of the mega oil fields, such as Ghawar in Saudi Arabia and Rumaila in Iraq (the elephants, as they are known in the business), have already been found. Even according to the USGS, global oil discovery peaked as far back as 1962. Today there are 1,500 large-scale oil fields known to exist around the world and they hold 94 percent of all crude oil. Four hundred of these contain nearly 70 percent of all the oil. And only forty-one of these have been found since 1980.

OPEC Rises Again

It should be said that figures like Campbell are in the minority. With their dire predictions of a global energy crunch, contemporary skeptics could join a pantheon of naysayers over the decades who prematurely forecast the demise of oil. Industry critics of Campbell and ODAC claim their research is skewed out of an environmental or anticorporate bias.

And yet, as writer Jeremy Rifkin points out in his book *The Hydrogen Economy,* the oil industry boosters and the peak production Cassandras are actually only arguing over a timeline. Neither side, it turns out, doubts that peak production is approaching. They just disagree on when it will happen. And whatever the timeline, one thing is certain—the last years of affordable world oil can come from only one place. That's the Middle East.

In total, the Middle East accounts for 65.3 percent or 685.6 thousand million barrels of proved oil reserves. Saudi

Arabia holds 26 percent of total reserves on its own. Equally important is an industry measurement called the reserve to production ratio, or R/P. In short, this looks at where a country's oil reserves are on Hubbert's bell curve. In the U.S., the R/P ratio is 10 to 1. That is, we can keep producing our nearly six million barrels a day for 10 more years at our current levels. But if you apply the same criteria to Saudi Arabia, its oil industry can keep pumping out eight million barrels a day for the next 55 years. Iran has 53 years at around 3.5 million barrels a day, and Iraq, which has seen its production levels reduced in recent decades by war and UN sanctions, has 526 years at its current levels (though this will fall as it ramps up production).

No other nations can come close to matching the reserves found in the Middle East. The rulers of the Arab oil nations and Iran will soon govern world oil prices and production once again. This time it will be for good.

To begin with, the Middle East's spare capacity and reserves will allow the OPEC nations to dictate the price of oil. Depending on the global political and economic climate, they may decide to keep prices relatively low in order to keep the developed world happy and the increasingly intertwined global economy in equilibrium. But OPEC has also shown its ability to use oil as a weapon. Once the Middle East regains its oil superiority, any production cut or new embargo aimed at the U.S.—to protest, say, the U.S.'s imperial aspirations or the continuing Israeli-Palestinian conflict—could send oil prices over $50 a barrel. Under this scenario, America's oil dependence could quickly shift the geopolitical balance of power firmly toward the Middle East. This wouldn't happen overnight, of course. The U.S. government, fearing threats to its hegemony, would obviously feel the need to use its might quickly by solidifying its influence on the governments of the major oil powers.

Yet with the oil stakes now so high, other new major consumers of oil, including Russia, China, and India along with the greater unified powers of Europe might feel the need to confront the U.S.'s control of the Middle East. It's a scenario that has played out many times before in the history of oil. But this time, with no new oil to alter things, the Middle East will be the endgame.

5

Energy Wars: Oil and National Security

Even if Saddam Hussein had wanted to placate the United States prior to the 2003 invasion of Iraq, the Bush administration had already made the decision to depose him.

The real reasons for invading Iraq were as simple as the stated reasons were obtuse and, despite the administration's unwillingness to declare them, they were entirely consistent with the principles that have guided American policy in the Middle East since the end of World War II. Saddam Hussein was not overthrown because he was helping Al Qaeda and not because he was developing weapons of mass destruction. No, whatever the Bush administration really knew about Saddam Hussein's military capabilities at the start of 2003, it believed that by removing him it could shape a U.S.-friendly regime in his place. In doing so, the U.S. would be able to command its own destiny in the one region that will dictate the fortunes of all the world's powers for decades to come.

The Middle East holds two-thirds of the world's known oil

reserves. Saudi Arabia and Iraq are home to the lion's share of that. The oil in those fields is cheap to extract—unlike fields in Russia, Mexico, and Canada, most Persian Gulf oil fields continue to operate below their full production levels. In the short term, consumers are able to draw on oil from all over the globe but as those other supplies dwindle or become too expensive to exploit, the major industrial nations will be forced once again to look to the Middle East.

It's a place where ethnic, religious, and small-town political rivalries meld together with a historic distrust of foreign intervention—hardly the type of environment any self-respecting superpower wants to be dependent on for its energy needs.

Of course the Middle East is not just America's problem. The whole world is dependent on oil and at least the U.S. remains a major producer of oil in its own right. But America's oil need is greater than any other nation.

A stable, i.e., U.S.-compliant, Middle East has been a national security goal of the United States for over three decades. That concern prompted President Jimmy Carter to issue what would become known as the Carter Doctrine—the willingness to intervene militarily against any threat to Middle East oil security. The same goal led the Reagan administration to fund and train Arab Islamic guerrillas, the mujahadeen, to counter the Soviet Union's invasion of Afghanistan, which the U.S. and Saudi Arabia were convinced was a Soviet plan to occupy the Persian Gulf. In 1991, George W. Bush's father put together a multinational coalition to repel Saddam Hussein from Kuwait and to protect Saudi Arabia after the dictator had made an audacious grab for Kuwait's oil fields. All of these events were reactions to a perceived threat to stability in the Middle East and American national security. And all were prompted by what the U.S. saw as aggressive action by either the Soviet Union or Iraq. The big difference in 2003 was that the U.S. didn't wait for an aggressor to strike first. The U.S. launched a preemptive attack on Iraq.

Oil addiction is America's Achilles' heel. To remain dominant in the world, the U.S. must be sure that oil flows freely and consistently onto the global market. The George W. Bush administration, with its decades of experience in the business and politics of oil, understands this better than most. And that is why it has made the protection of global oil supplies an equal partner with the war on terrorism in guiding U.S. foreign and defense policy.

The war to rid Iraq of Saddam Hussein was just the curtain-raiser for an ambitious and risky new foreign policy to solidify America's preeminence. It revolves around shoring up and establishing U.S.-friendly regimes all over the world to ensure the unfettered flow of cheap oil. And it dovetails neatly with the global war on terrorism because so much of that fight takes place in the Middle East, the Caucasus, Latin America, and south and central Asia, all flashpoints of terrorist activity and all loaded with oil.

This is not to suggest that the U.S. will send in the cavalry to shore up every leaky pipeline from Aceh to Turkmenistan. But, using the justification that economic security is part of national security, American armed forces will, when necessary, back up a strategy to protect the secure supply of oil. This new foreign policy is a far cry from the humble tone set by the Bush administration during its first days in office. In many ways, it can be seen as a direct reaction to the terrorist attacks of September 11, 2001. To use the sporting metaphor, the Bush administration decided that the best defense is a good offense.

Smoke and Mirrors

Historians will no doubt record—once the fog of war and the smog belched by politicians clear—that the justification for the Iraq war was driven by subterfuge as great as the Gulf of Tonkin incident that led to the escalation of the Vietnam War or the explosion aboard the USS *Maine* that proved the catalyst for the 1898 Spanish-American War. And all because the antagonists were afraid to call the war for what it was.

Both George W. Bush and British Prime Minister Tony Blair, with varying degrees of eloquence, insisted theirs was a moral campaign to rid the world of a tyrant who posed an imminent and deadly threat both to his neighbors and to the rest of the world. The U.S. continually cited Saddam's ties to the Al Qaeda terrorism network, as well as an active chemical and biological weapons program that he had used before on his neighbor Iran and on his own people. The one thing the administration refused to talk about was oil.

"It has nothing to do with oil, literally nothing to do with oil," Defense Secretary Donald Rumsfeld told *60 Minutes* with steel-jawed condescension in the fall of 2002.

But, at demonstrations that drew millions of people all over the world, protestors shouted, "No Blood for Oil"—a peace mantra accusing the U.S. of a simple smash and grab to take control of Iraq's vast reserves of oil.

Certainly, history seemed to be on the side of the demonstrators. Western governments and oil companies have a long history of meddling in the internal affairs of Middle East states. There was the CIA overthrow of Iranian president Mohammed Mossadegh that suppressed Iran's fledgling attempts at nationalizing its oil industry. That opened the way for U.S. oil companies to negotiate lucrative deals with the new U.S.-installed Iranian ruler, the Shah. And there was the almost symbiotic manner in which the U.S. government and U.S. oil companies handled their affairs with Saudi Arabia in the years following World War II. Often during that time, the oil companies acted as the government's eyes and ears in the region. Surely then, the Bush administration was willing to overthrow Saddam Hussein as a favor to its friends in Big Oil.

But to those who understood the modern business of oil, this simplistic if catchy argument didn't quite make sense. For one thing, the U.S. has gone to great lengths to diversify its oil imports in recent years. In 2003, it received most of its foreign oil from Canada, Mexico, and Russia—only 3 percent of its oil needs are met by Iraq. Putting U.S. companies in charge of Iraq's oil infrastructure would certainly guarantee more of that oil came to the U.S., but controlling oil fields doesn't guarantee America's energy security. Iraq's oil, like all crude oil, flows onto one global market. U.S. oil companies and independent refiners pay the price the global market dictates wherever they get their oil from.

It's the price, not the source, of oil that protects or threatens America. And it was Saddam's threat to oil prices that most worried the U.S.

In 1980, Saddam Hussein had invaded Iran, attacking its oil fields and sending oil prices skyrocketing to their highest levels ever. During the Gulf War, Saddam Hussein also attacked Israel and Saudi Arabia with ballistic missiles. The Gulf War sent oil prices above $40 a barrel, the highest they had been since the outbreak of the Iraq-Iran war in 1980. The economic fallout that followed sent the United States into recession for two years—and with it, George Bush's hopes of reelection.

Then, in April 2002, Iraq announced it was withholding its oil from the market for a month to spur a boycott by all Arab producers in support of the Palestinians. Iraq's attempt to use oil as an economic weapon failed miserably, but it only underscored how Saddam Hussein remained a threat to oil security. Oil traders started factoring in a premium to the price of oil to take into account uncertainty in the market. The U.S. didn't invade Iraq just to grab its oil. It invaded so that it could dictate the long-term political future of the Middle East and guarantee a steady and stable supply of Middle East oil for decades to come. That meant getting Saddam Hussein out of the picture.

Seeds of War

Neoconservative factions in the Republican Party had been calling for the overthrow of Saddam Hussein ever since the end of the first Gulf War. Many believed that Bush the Elder and Colin Powell, then chairman of the Joint Chiefs of Staff, had made a crucial mistake in not marching on Baghdad and finishing off Saddam Hussein. By failing to finish the fight, they argued, the U.S. had reinforced the Vietnam-era perception, hammered home by the 1983 bombing of the US embassy in Lebanon and the subsequent withdrawal of U.S. forces from the country, that the superpower didn't have the stomach for a fight. Saddam Hussein was also viewed as the most aggressive regional threat to Israel, a nation neoconservatives had long supported. Removing him, it was argued, could also undermine the regional strength of Syria, Israel's most powerful enemy after Iraq. And all of that would improve the prospect for a long-term peace in the Middle East.

In 1998, a conservative think-tank called the Project for the New American Century wrote a letter to President Bill Clinton urging him to depose Saddam Hussein. The think-tank's founding principles were "to shape a new century favorable to American principles and future challenges; a foreign policy that boldly and purposefully promotes American principles abroad; and national leadership that accepts the United States' global responsibilities."

The letter to Clinton amplified those principles. If Saddam Hussein was not overthrown, it said, "the safety of American troops in the region, of our friends and allies like Israel and the moderate Arab states, and a significant portion of the world's oil supply will all be put at hazard." The letter also implored Clinton to go it alone and avoid having U.S. foreign policy "crippled by a misguided insistence on unanimity in the UN Security Council."

It was signed by eighteen national security hawks including the future Vice President Cheney, Defense Secretary Rumsfeld, Deputy Secretary for Defense Paul Wolfowitz, Assistant Secretary of State Richard Armitage, and Elliot Abrams, the National Security Council's top Middle East official.

A report produced by the think-tank in 2000 took a more expansive worldview. It called for the installation of a "substantial American force presence in the Gulf," so that it could maintain "global U.S. pre-eminence."

Upon becoming president in 2000, George W. Bush asked Vice President Dick Cheney to convene a task force to plan a new energy strategy for the nation. The National Energy Policy, released in May 2001, identified an impending energy crisis. "On our present course," said the report, "America twenty years from now will import nearly two out of every three barrels of oil—a condition of increased dependency on foreign powers that do not always have America's interests at heart." The answer, said the National Energy Policy, was to increase domestic oil production and to make "energy security a priority of our trade and foreign policy."

*   *   *

Immediately after September 11, Deputy Secretary of Defense Paul Wolfowitz began campaigning for the overthrow of Saddam Hussein. Wolfowitz, who would be credited as the architect of the plan to overthrow Saddam Hussein, believed that remaking Iraq as a Western-style democracy would spur similar democratization throughout the Middle East. In a Saddam-free Iraq, the U.S. would have the sort of dependable ally it had been lacking since the overthrow of the Shah of Iran in 1979.

In August 2002, Cheney gave a speech to veterans in Nashville, Tennessee, where he telegraphed the Bush administration's decision to overthrow Saddam Hussein. If the Iraqi leader acquired weapons of mass destruction, said the vice president, he would "seek domination of the entire Middle East" and he would "take control of a great portion of the world's energy supplies."

Still, in the months leading up to the war, the administration downplayed the themes Cheney had sounded. Instead, a phalanx of Bush officials talked incessantly of Saddam Hussein's ties to Al Qaeda and of weapons of mass destruction. Not only did Saddam threaten his only people, they said, but he was also trying to build a nuclear weapon.

The official Iraq war was over in under a month. Saddam Hussein's regime crumbled almost too easily and the dictator himself disappeared without a trace. President Bush announced the end of major hostilities from the flight deck of the *Abraham Lincoln* aircraft carrier. It was a gaudy and grandiose piece of reelection PR that he soon came to regret, for immediately

U.S. forces found themselves sitting targets of insurgents egged on by taped messages from Saddam Hussein himself. As the summer of 2003 turned into fall, and as U.S. soldiers fell at the rate of one a day, it became impossible for the administration to dismiss the organized resistance.

Having been promised by the president that the U.S. would fight a decisive campaign followed by a short transition to civilian Iraqi power, Americans began to realize that the military faced a long-drawn-out guerrilla war.

It was hardly the only area where the U.S. public was misled. In July 2003, the Bush administration was forced to admit that it had no real basis for its claims that Iraq was developing nuclear weapons. Neither did the administration have any new evidence of Iraq's chemical and biological weapons program aside from what it had received from the UN in the 1990s.

Saddam Hussein was finally captured by the U.S. in December 2003, but still the administration was unable to demonstrate any links between the dictator and Al Qaeda. The Bush administration was able to show the gruesome mass graves of Saddam Hussein's torture victims, but genocide was not the factor that had propelled the U.S. and UK into invading Iraq.

What's more, as students of Middle East affairs were quick to point out, Saddam's genocidal tendencies dated back to the early 1980s—when certain members of the current Bush administration actively courted Saddam. The diplomacy conducted back then by the Reagan administration paid scant attention to biological weapons and war crimes. The administration had more important matters of realpolitik to contend with. Saddam was fighting newly revolutionary Iran, so the U.S. viewed him as a natural ally. The man the Reagan administration sent to negotiate with Saddam was a young Donald Rumsfeld.

Central to Rumsfeld's discussions were Iraq's oil supplies. In 1982, Syria, which backed Iran in its war with Iraq, had shut down the Trans-Syrian oil pipeline, which had carried 400,000 barrels a day. Almost immediately, San Francisco–based contractor Bechtel began planning a new pipeline for Iraq's oil that would flow out of the country to the Gulf of Aqaba on Jordan's Red Sea coast. The pipeline would bypass the southern Straits of Hormuz, a treacherous maritime passageway that Iran had turned into a war zone.

Bechtel's former CEO was the then Secretary of State George Schultz and in late 1983, a top Bechtel employee was invited to meet with the State Department to discuss the pipeline plan. Later that month, Reagan sent special envoy Rumsfeld to Baghdad for a meeting with Tariq Aziz, Iraq's deputy prime minister, and Saddam Hussein. Rumsfeld's mission was to inform Iraq that the U.S. was willing to help build the Bechtel pipeline so it could export more oil to the West.

Rumsfeld's overtures to Iraq came at a most unfortunate time. Just two months before, Iran had reported to the United Nations that Iraq had attacked both its troops and civilians with chemical weapons. On March 26, 1984, the same day that

Rumsfeld returned to Baghdad to meet again with Aziz, a UN team independently confirmed that Iraq was using chemical weapons. The attacks embarrassed the Reagan administration but didn't stop its dialogue with the Saddam Hussein regime. In April, U.S. diplomats asked the Iraqis not to purchase chemical weapons from U.S. suppliers. "We would ask the GOI [government of Iraq's] cooperation in avoiding situations that would lead to a difficult and potentially embarrassing situation," was how Secretary of State Schultz put it in a memo.

Even after the U.S. officially condemned Iraq for its chemical weapons attacks, the State Department asked the U.S. Export-Import Bank to approve short-term loans for Iraq to help build the pipeline. But in the end the project fell through and relations between the U.S. and the Hussein government stagnated.

It would appear now that the U.S. decided to overthrow Saddam Hussein when it did because it could. Following the end of the first Gulf War, the U.S. military routinely attacked Iraq's armed forces in the two no-fly zones that were established by the UN to protect Iraq's Kurdish and Shiite populations in the north and south of the country.

Frankly, given the damage the United States had inflicted on his military infrastructure, Saddam Hussein posed no threat and wasn't in much of a position to threaten his neighbors, either. It seems certain that he didn't have nuclear weapons and his army, though still large, was a shadow of the force that invaded Kuwait in 1991.

What to Do with Iraq's Oil

The first thing the U.S. military did in the second war with Iraq, even before the first smart bombs fell on Saddam's "bunker," was to send in special forces to secure Iraq's northern and southern oil fields. And as U.S. marines and the Third Infantry marched into Baghdad, they made sure that Iraq's oil ministry alone was protected even as looters ransacked all other Iraqi government offices, the city's hospitals, and even the national museum.

Oil may not have been the only reason for going to war, but it became an overriding factor in deciding Iraq's future. Oil, said the administration, would form the bedrock of a new Iraqi society. It would pay for the rebuilding of Iraq's infrastructure and it would finance a new and unique Middle East democracy.

Iraq has listed proven reserves of 112.5 billion barrels, about 11 percent of the world's total. But little new exploration has been undertaken in Iraq since the oil industry was nationalized in 1972 and the EIA estimates Iraqi reserves could total over 400 billion barrels. Even accounting for the penchant of Middle East nations for inflating the quantity of their oil reserves, there is no denying that Iraq has a mother lode of oil, the majority of which sits just 600 meters below the surface and needs no pumping—it is so plentiful that it gushes up under its own pressure. That makes Iraq oil very cheap to produce—as little as $1.50 a barrel—and offers huge profit margins when compared to production costs that range from $6 to $15 a barrel in other parts of the world.

But will oil help rebuild Iraq or, like so many oil-rich nations before it, will its total dependence on oil revenues simply accentuate the divisions that already threaten to tear the country apart?

One World Bank report published in 2000 on the economics of conflict looked at 160 countries between 1960 and 1999 and found that the more a country is dependent on natural resources, the higher the risk it will fall into civil war over the next five years. When oil was the primary resource, the risk was 40 percent more likely.

Other political scientists suggest that even when oil is not itself the sole catalyst for conflict, the potential to control a state's oil revenues can intensify fighting and when one side has control of those resources, revenue from that oil can prolong civil conflict by financing the war effort. Civil wars in Angola, Colombia, and Sudan, for example, all erupted long before the oil potential of those countries was fully realized, yet control or sabotage of the oil apparatus became an important aspect of each civil war. If America needed a reminder of this, Iraq's main oil pipelines were repeatedly attacked by insurgents.

Iraq faced another major problem. Since the British cobbled together modern Iraq out of the remnants of the Ottoman Empire at the end of World War I, the country has been a tinderbox of balkanized ethnic groups, including Sunni and Shia Muslims who are distrustful of each other, the Kurds in the north who are distrustful of both Muslim groups, and the ethnic Turks who fear the Kurds' ambitions for an independent state. The bulk of Iraq's oil reserves sit in the Kurdish region around Mosul in the north and in Shia regions to the south, including the vast Rumalia oil fields. Both the Kurds and Shia were persecuted by Saddam Hussein, but it was exactly this ruthlessness that held the fractious and artificial makeup of Iraq together. Iraq, bereft of its autocratic leadership, was in danger of splitting along strict ethnic and religious lines. The struggle to control Iraq's oil would only compound those divisions.

Then there was the perception that the United States just wanted the oil for itself. Back in December 2002, a joint study by the Council on Foreign Relations and the James A. Baker III Institute for Public Policy of Rice University recommended the best way for the U.S. to use oil to help the Iraqi people and not be perceived as an imperialist force. The report's main recommendations were that a) Iraqis, not the U.S., should maintain control of their own oil sector; b) a large portion of early oil revenues should be spent on rehabilitation of the oil industry; c) there should be "a level playing field" for all international oil production and service companies to participate in the new Iraq; and d) oil revenues should be fairly shared by all of Iraq's citizens to avoid the social strife that affects so many oil-rich nations. Otherwise, warned the report, the U.S. would quickly alienate the Iraqi people and be seen as an imperialist force.

The Bush administration, however, seemed to be going out of its way to create the opposite perception. As early as

November 2002, the Pentagon had offered Kellogg Brown and Root, the oil-services arm of Dick Cheney's former company Halliburton, a secret no-bid contract to extinguish oil fires and get Iraq's oil industry back on line in the event of war. Originally worth $1.4 billion, it quickly ballooned to over $2 billion as the problems of rebuilding Iraq's infrastructure became clear.

In recent years, the UN Oil-for-Food program worked to ensure that any oil Iraq sold on the global market would be used to supply food for its people. Advocates of using oil to rebuild Iraq often touted the program as a transparent way of ensuring the Iraqi people continued to benefit from oil revenues. But, in the summer of 2003, the UN, under pressure from the U.S., transferred $1 billion from the Oil-for-Food program into a replacement program, the Development Fund for Iraq. This fund is controlled by the U.S. and advised by the World Bank and International Monetary Fund. All future proceeds from the sale of Iraqi oil will be placed into the fund.

Rather than directly benefiting the people of Iraq, the new fund will more likely be used to attract new investment into Iraq's oil industry. Some of the funds, according to a study by the Sustainable Energy and Economy Network, a DC-based nonprofit research group, will be used as collateral for projects approved by the U.S. Export-Import Bank, which helps promote U.S. business interests abroad. The role of the bank shouldn't be underestimated. Many U.S. companies are eager to get into Iraq but have been unable to get private bank credit because of the risk involved. Iraq's oil money, it seems, will remove that risk.

Rocking the Region

In time, and with a great deal of Western investment, Iraq will once again be a major oil power. But this would cause a great deal of unease among the other major Middle East oil producers—not least Saudi Arabia.

Conservatives in and out of the Bush administration salivate at the idea of using Iraqi oil as a weapon to destroy OPEC. It was OPEC, after all, that humbled the U.S. in the 1970s and the majority of its members are virulently opposed to America's main Middle East ally, Israel. Furthermore, OPEC is destined to be the most important force in world oil once again. According to the U.S. Department of Energy forecast, OPEC nations will meet 62 percent of all world oil demand by 2020.

By withdrawing from OPEC, Iraq would be free of the production quotas the cartel sets to maintain its $25 average price per barrel. Iraq, fueled by Western oil company investment, could then ramp up its own production, causing oil prices to drop and undercutting the profit margin of OPEC members. Faced with such price pressure, the thinking goes, OPEC would crumble. Only Saudi Arabia would be able to pump enough oil to keep paying its bills at the lower price.

Ultimately, even Saudi Arabia could suffer as a result of Iraq's oil production. Saudi Arabia has domestic problems that make it particularly vulnerable to fluctuating oil prices, not to mention the growth of a fledgling democracy on its borders. Saudi Arabia has the largest reserves of oil in the world but its total reliance on oil profits left it mired in debt when global oil prices plummeted in the 1980s. Oil revenues account for 40 percent of the kingdom's GDP and nearly 80 percent of all state revenue. Taxation is minimal; instead, government national spending on state services—water, health, education, telecommunications, and the airline—adds up to 24 percent of Saudi Arabia's GDP. That sounds pretty good until you take into account that, for all that oil wealth, unemployment runs at nearly 20 percent. Quality of life has fallen considerably in what is essentially an oil-fed welfare state and as people's lives have become harder, an increasingly restless majority underclass, many of whom identified with the country's ultraconservative Wahhabi Islamic clerics, has put pressure on the autocratic regime of the ruling sheiks.

The Saudi monarchy has long been happy to tolerate anti-American extremism as a way of deflecting dissent away from its own problems at home. Fifteen of the nineteen Al Qaeda terrorists that carried out the September 11 attacks came from Saudi Arabia and the monarchy makes no attempt to hide its funding of militant Islamic and anti-zionist causes. But Saudi Arabia's domestic anti-Western strategy no longer seems to be placating the more radical Islamic factions. Recent terrorist attacks in the kingdom have focused on mainly Saudi targets and seem intent on destabilizing the monarchy.

Which is the last thing the U.S. wants to see, whatever the

Bush administration says about promoting democracy in the region. The U.S. and the Saudis have looked out for each other ever since the end of World War II. The Saudis make sure oil keeps flowing to the West at a stable price; the U.S. ensures Saudi Arabia's regional security. During the 1991 Gulf War, the Saudis upped production to counter the loss of Kuwait's daily oil production. In 2003, they stepped up production to make up for loss of Venezuelan oil during its month-long general strike and to cover the shortfall when Nigeria's oil fields shut down during ethnic clashes and a general strike.

Whatever the Bush administration feels about Saudi Arabia's commitment to purging Al Qaeda from its land, and however much it might like to see the breakup of OPEC, the autocratic Saudi regime remains an important component, along with Kuwait and now Iraq, in cementing U.S. influence over the major Middle East oil producers. America, along with the rest of the developed world, will need these nations in the decades to come.

The Post-Iraq World

In the summer of 2003, only a few weeks after the end of the official hostilities in Iraq, the Pentagon unveiled the biggest reassignment of its forces worldwide since the end of the Cold War. Defense Secretary Donald Rumsfeld had been pushing for a shake-up of the armed forces since his first days in office, advocating a more mobile fighting force that could react to hot spots all over the world. At first he had butted heads with traditionalists who believed in the need for a large standing army, but his thinking gained new credence in the months following September 11, 2001, as the U.S. launched a rapid-response war on terrorism in many parts of the world.

With most of the former Eastern bloc countries clamoring to become part of NATO, Rumsfeld, along with many other military observers, questioned the logic of having 70,000 U.S. troops stationed in Western Europe. And with many other small nations more than willing to tolerate U.S. bases in return for substantial bundles of cash, the U.S. took the opportunity to reassign its troops to the areas of the world it now believed were of most strategic importance.

The Pentagon announced it would reduce the large garrison forces in the U.S., Germany, and South Korea, which together hosted 80 percent of America's 1.4 million troops. Instead, the U.S. would rotate its forces through a large number of bases scattered all over the world, with special attention given to the so-called "arc of instability" running through the Caribbean rim, Africa, Central Asia, the Middle East, South Asia, North Korea, and the Caucasus. The new formations would include regional hubs as well as forward-operating bases that might house just a few dozen troops but could be quickly transformed into action-ready staging posts.

One thrust of the military shake-up was to make U.S. forces better able to target suspected centers of terrorism. The second reason was to protect oil supplies. As Gen. Charles Wald, deputy commander of the U.S. European Command, explained to the *Wall Street Journal*, "In the Caspian you have large mineral reserves. . . . We want to be able to assure the long-term viability of those resources."

One look at a map of the new troop deployment told the story. With new military bases in Romania, Bulgaria, Azerbaijan, Kazakhstan, Kyrgyzstan, and Uzbekistan the U.S. could keep watch over the Caspian and the all-important Baku-Ceyhan oil pipeline; other bases in Afghanistan, Qatar, Saudi Arabia, Djibouti, and Oman (not to mention Iraq) guaranteed a strong presence in the Persian Gulf.

The first signs of this new oil-protection policy had become apparent the summer before. Congress had authorized $98 million for U.S. troops and equipment to help the Colombian army to protect oil pipelines owned by the California company Occidental. The pipelines were regular targets of the FARC and ELN, the two main leftist rebel groups in Colombia's decades-old civil war. In the spring of 2003, just as U.S. forces were invading Iraq, a far smaller group of seventy Green Berets flew into Colombia to secure Oxy's pipeline.

The funds were authorized under the proviso of the administration's war on terrorism, but the military training had more to do with the National Energy Policy. The Andean nations of Ecuador, Venezuela, and Colombia contribute 20 percent of the U.S.'s imported oil. Colombia is the tenth-largest oil supplier for the U.S. The Bush administration has made increased imports from those nations an important part of its goal of lessening its dependence on Middle East oil. Colombia's oil is easy to produce and output could be significantly increased were the oil companies not targeted so often. The national government also uses a good deal of its oil profits—25 percent of the country's annual revenues—to fight the rebels.

U.S. troops have operated as advisers in Colombia for decades, but the move to train Colombian troops raised the profile of U.S. military involvement in one of the world's longest and most treacherous civil wars. It is a conflict that shows no sign of abating and one that could suck in more troops in the future.

Cutting Deals in the Caspian

Colombia seems relatively simple compared to the Caspian region. Since the breakup of the Soviet Union and the revitalization of the Russian oil industry, the Caspian Sea has been the scene of frenetic business activity and political maneuvering as oil companies from all over the world have worked to secure potentially lucrative production agreements with the independent states of the Caspian region: Azerbaijan, Kazakhstan, Turkmenistan, and Uzbekistan.

No nation has shown more interest in Caspian oil than the United States. Immediately following the fall of the Soviet Union, the U.S. quickly grasped that Caspian resources could provide an important alternative to Persian Gulf oil. Working hand in hand with U.S. oil companies, successive administrations have tried to influence the region. As Anna Borg, a deputy assistant secretary for energy issues at the State Department told Congress in 2003, "The U.S. government and State Department are focusing on [the Caspian] extensively."

The Caspian Sea could prove to be one of the largest oil finds in the world after the Middle East. Or, more likely, it could hold much smaller, though still important, reserves, along the lines of the North Sea or Alaska's Prudhoe Bay. Estimates vary from 17 billion barrels of proven reserves to 70 billion depending on whether you talk to Caspian oil boosters or naysayers.

During the 1990s, the various components of ExxonMobil, BP, ChevronTexaco, and TotalFinaElf, along with many smaller oil companies, courted the republics that bordered the Caspian Sea to secure drilling concessions on what everyone hoped would be a grand prize. No nation received more attention during this time than Azerbaijan, whose leader—the canny former Soviet strongman, Haydar Aliev—quickly aligned himself with the U.S. by portraying his nation as a crucial buffer to Iranian-backed Islamic radicalism.

It also didn't hurt that Azerbaijan had claims on what then appeared to be huge oil reserves. In short order, a little-known outfit called the U.S.-Azerbaijan Chamber of Commerce became one of the most powerful lobbying outfits on Capitol Hill.

The Chamber of Commerce board of directors and advisers reads like a who's who of the Bush, Reagan, and Carter administration heavy hitters. Over the years, it has included former Bush Secretary of State James Baker, former Bush National Security Adviser Brent Scowcroft, former Texas Senator Lloyd Bentsen, former Nixon and Ford Secretary of State Henry Kissinger, and members of the George W. Bush administration including Vice President Dick Cheney,

Assistant Secretary of State Richard Armitage, and former head of the Defense Policy Board Richard Perle.

While Azerbaijan was assembling a lobbying A team, other Caspian countries like Kazakhstan and Turkmenistan were accepting offers on other Caspian fields. Chevron, for example, took a 50 percent stake in Tengiz (the Kazakh word for sea) with the Kazakhstan national oil company in return for a promise to invest $20 billion in the field over forty years. None of the major oil companies wanted to be left out of the Caspian game and so they were soon cutting deals with some of the most unsavory regimes in the world—all hangovers from the Soviet totalitarian system and all hell-bent on making a bundle in petrodollars from the Westerners.

U.S. companies like Mobil and Exxon and the British and French giants BP and Total were eagerly egged on in their endeavors by their own governments, while Russia sat, emasculated, watching on the sidelines. Meanwhile the appalling human rights record of these regimes was conveniently ignored. Both Presidents George H.W. Bush and Bill Clinton took regular trips to the region during their time in office, glad-handing with old-school autocrats like Kazakhstan's President Nursultan Nazarbayev and Turkmenistan's President Saparmurat Niyazov.

That Caspian oil could cure Western dependence on the Middle East was accepted without question. Before long, the question for the major powers was not how they could secure the oil but how they could transport it out. The oil companies, and the governments that backed them, now began to hatch different competing plans to build a pipeline that would carry Caspian oil to market. By 1995, industry maps of the Caspian looked like a spaghetti junction of rival pipeline bids—one heading west through Georgia and Turkey, one going north through Russia, another pointing east via China and yet another cutting south via Iran to the ports of the Persian Gulf.

At times, the geopoliticking that surrounded the Caspian pipeline projects was high farce. Ideally, Kazakhstan should have simply exported its oil via the same Russian pipelines it had used during the old Soviet days. But Russia was angry that its former vassal states were allowing the U.S. a grab at what it believed was its own sovereign oil. Russia's revenge was to shut down the pipelines that Kazakhstan depended on to move oil to market.

All of the Western oil companies saw a quick and straightforward solution to exporting the Caspian's landlocked oil. That was to build a short and inexpensive pipeline south through Iran to the Persian Gulf. But that involved doing business with the Iranian government, which was illegal for U.S. companies. So incensed were U.S. oil companies at having their hands tied that Dick Cheney, then chairman of Halliburton, called on the Clinton government to lift the Iran ban. The Clinton administration understood all too well the importance of Caspian oil and toyed with building diplomatic bridges with Iran, but the ban remained.

At one point, the California company Unocal began negotiations with the Taliban regime of Afghanistan to build a pipeline as it searched for a way to export natural gas from

Turkmenistan to Southeast Asia. The talks failed and the Taliban would soon be overthrown but at least one member of the Unocal enterprise came out on top—Hamid Karzai, the man the U.S. installed to lead Afghanistan after the Taliban's demise, was Unocal's lead negotiator on the pipeline project.

With Russia uncooperative, Afghanistan too unstable, and Iran considered taboo, the Clinton administration finally threw its support behind a million-barrel-a-day pipeline project that would run west from Baku, through Nagorno-Karabakh, a region claimed by both Azerbaijan and Armenia, into Georgia and onto Turkey, where it would emerge at the Black Sea port of Ceyhan. From there, oil would be shipped via tanker through the Bosphorus and into the Mediterranean on its way to Europe and the U.S.

Most independent observers believe the Caspian Pipeline Consortium, as it is known, makes little logistical sense. Not only must the pipeline traverse a smorgasbord of ethnic unrest but it will result in millions of barrels of oil being shipped through the Bosphorus, already one of the world's most precarious and congested shipping lanes. For a while it seemed unlikely that the pipeline would be built, but following BP's purchase of U.S. oil company Amoco in 1998, the British giant gave in to pressure from the Clinton administration and agreed to take a 34 percent lead in the CPC plan. Ground was broken on the CPC project in 2002 and the first oil is expected to flow in 2005.

Caspian oil is an important part of America's overall oil diversification strategy and one the Bush administration believes is well worth protecting.

As the U.S. military completes its redeployment to more strategically important parts of the world, the Caspian will be in easy reach of American forces. With new military bases in Romania and Bulgaria, a U.S. rapid response force is in easy reach of the Caucasus region to counter any threat to oil supplies that might come from Turkey, Iraq, Chechnya, or even Iran. At the same time, U.S. forces have trained the Georgian military to counter armed Islamic groups operating out of the lawless Pankisi gorge. The stated purpose of this training is to fight the War on Terror, but as we've seen these oil and terror priorities dovetail before, Georgia is also an important part of the CPC pipeline route. Meanwhile, on the Caspian's eastern flank, it seems unlikely that the U.S. will abandon the new military bases it established in Uzbekistan and Kyrgyzstan during the Afghan war, despite Russian President Putin's reservations.

The idea of another great game may have been overblown but there is plenty of geopolitical maneuvering still taking place in the region. Recently China's national oil company cut a deal with Turkmenistan to produce and ship gas to the rapidly growing Chinese market. Russia is also intent on pumping as much oil and gas as it can to sustain economic growth. Who knows, Afghanistan might yet host a pipeline of its own. Stranger things have happened in the oil industry and certainly in this particular part of the world.

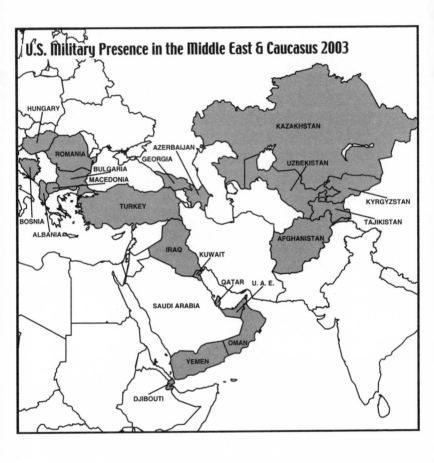

Kuwait of Africa

In June 2003, on the eve of President Bush's first visit to Africa—only the second time a sitting president had visited the continent—Gen. James Jones, commander of the U.S. European command, gave an interview to the *New York Times* in which he outlined another layer of the U.S. military's shift in global priorities.

The U.S. was negotiating with a number of African nations for the long-term use of military bases to help combat the terrorist groups that may be operating in the region. The areas of interest included Algeria, Morocco, and also sub-Saharan venues like Mali. Augmenting these bases would be a strong U.S. navy and marine force operating in the Gulf of Guinea. "The carrier battle groups of the future may not spend six months in the Mediterranean sea," Jones said, "but I'll bet they spend half the time going down the west coast of Africa."

There are very few known terrorist outfits in the waters off west Africa but there is an enormous amount of oil. Africa possesses an estimated eighty billion barrels of oil, 8 percent of total world crude reserves. That's not enough to shake the Middle East's hegemony but it does make Africa a very desirable part of the U.S.'s pursuit of foreign oil diversity. In 2003, sub-Saharan Africa was pumping four million barrels a day, most of it from countries on the Gulf of Guinea. That's as much as Iran, Venezuela, and Mexico combined. The U.S. imports some 16 percent of its foreign oil from this part of the world and by 2015 it is expected to rise to 25 percent. And that makes African oil, in the words of U.S. Undersecretary of State for African Affairs Walter Kansteiner, "a national strategic interest."

As the African Oil Policy Group, a Washington lobbying group, reported to the House of Representatives African subcommittee in 2002, "the Gulf of Guinea oil basin in west Africa, with greater western and southern Africa and its attendant market of 250 million people located astride key sea lanes of communication, [is] a vital interest in U.S. national security calculations."

An estimated 24 billion barrels of oil sit in the Gulf of Guinea. This oil is attractive to both the U.S. government and the major oil companies for a number of reasons. For one thing, it is a light crude that is easily refined into gasoline. Then there is the close proximity to the U.S. East Coast, cutting down on transportation costs. But, aside from the inconvenience of having to drill offshore, this new west African oil also comes free of baggage associated with oil production in the developing world. Because of its isolated location, the oil companies are unlikely to run foul of competing ethnic groups or other citizens unhappy that they are not benefiting from oil revenues flowing to their governments. And aside from Nigeria, none of these new oil powers are members of OPEC and the U.S. intends to keep it that way.

Nigeria, Angola, and Equitorial Guinea, all of them already serious oil powers in their own rights, have territorial claims to the Gulf of Guinea and U.S. oil companies have been quick to snap up profit-sharing agreements (PSAs) and direct drilling concessions with these often troubled and corrupt nations.

But it is the tiny island republic of São Tomé and Principe, a former Portuguese colony located some 500 miles west of Equatorial Guinea, that has really whetted the appetite of the players in this latest oil game. Over twenty oil companies, including ChevronTexaco and ExxonMobil, have bid for the right to drill in the waters off São Tomé in a concession that is jointly run by the island's government and neighboring Nigeria. Under the terms of a 2001 accord, Nigeria will get 60 percent of the royalties on agreed contracts and São Tomé the rest.

São Tomé has also been wooed by the U.S. navy which would like to build a base on the island. Indeed, for a leader of just 150,000 people, President Fradique de Menezes could be said to be hitting above his weight. In September 2002, de Menezes was one of nine African leaders that Bush hosted at the White House. All nine preside over sizeable deposits of African oil.

The whiff of oil riches has already changed life on São Tomé. In July 2003, the de Menezes government was briefly toppled in a military coup by officers unhappy at their cut of future oil revenues. De Menezes later placated the army chiefs and was reinstated. The coup was just the first taste for São Tomé of the upheaval oil riches could bring. To get a fuller picture, it need only look at the experience of Equatorial Guinea a few hundred miles away.

In 2003, Equatorial Guinea produced 181,000 barrels of oil a day, not a huge amount by OPEC standards but enough to show this small west African nation has real potential. By 2005, petroleum output is expected to grow by 140 percent. U.S. oil companies, including ExxonMobil and Chevron, have invested $5 billion in the country's oil infrastructure and the country has been quite magnanimous in return: the oil companies get to keep 75 percent of all the oil receipts. Since the oil boom began in 1996 Equatorial Guinea's economy has grown 40 percent annually but its two traditional agricultural industries, cocoa and snail farming, have been decimated as workers have abandoned these pursuits in a chase for oil riches. The government of President Teodoro Obiang Nguema Mbasogo, who came to power by executing his uncle in a 1979 coup (perhaps not the best omen for the country), runs an efficiently autocratic oil regime, jailing opposition leaders and thumbing its nose at Western entreaties that Obiang make the country's oil deals more transparent.

Under the Clinton administration, Equatorial Guinea's oil potential was steadily cultivated but the government held back from offering a full embrace because of Obiang's dismal human rights record. But as the Bush administration sought new alternatives to Middle East oil, it came to view Equatorial Guinea in a new light.

As an official of one U.S. oil company operating in the country told the *Nation* in 2002, "For a long time our relationship with Equatorial Guinea revolved around human rights. That's a legitimate concern, but now that the energy picture is changing, that introduces something to balance out the dialogue."

Like many other times in the past, U.S. oil companies are acting as the eyes and ears for the American government in Equatorial Guinea. It also doesn't hurt that many of these companies have strong ties to the Bush administration. The CEO of one firm, CMS Energy (which later sold its holdings to Marathon Oil), contributed $100,000 to the Bush–Cheney 2001 Presidential Inaugural Committee. Tom Hicks, the chairman of Triton (another U.S. company active in Equatorial Guinea), is the fourth-largest contributor to Bush's political campaigns. And when Hicks bought the Texas Rangers in 1998, he helped make George W. Bush a millionaire many times over in the process.

The Implications of Our Energy Security

The terrorist attacks of September 11 made the Bush administration, and George W. Bush in particular, take a new hard look at the world. What they saw, in varying degrees, was a melting pot of resentment and anger directed by much of the developing world at the rich nations of the First World and the U.S. in particular.

A lot of that resentment grew from a decade of globalization that major world leaders and global economic institutions had promised would carry the whole world forward but which, instead, had left much of the earth's population behind while a small cadre of already wealthy nations profited. Head of the pack was the U.S., the world's only superpower, whose economic clout, military strength, and cultural hold over global society, be it through the McDonaldization of cuisine or the Hollywoodization of cultural values, generated awe, envy, and often outright hatred in many parts of the world.

In digesting September 11, the Bush administration had the opportunity to fashion its response in two very different ways. The first was to embrace, as Bush himself put it in his 2000 inauguration speech, a "humble" approach to the other nations of the world. Yes, the United States had been attacked on its own soil for the first time since Pearl Harbor but that didn't mean America was suddenly weak. It remained the most important and strongest nation on earth. It had the chance to appreciate that strength and see how others less fortunate might both sometimes unjustly resent the United States but sometimes have real grievances about how it conducts its global affairs.

The second was to close ranks and minds and, scared of another attack, consider the rest of the world a potential enemy. This is the course George W. Bush chose to adopt. "You're either with us or against us," became not just a rallying cry for the War on Terror but for the administration's approach to the rest of the world. There was no room for introspection or tolerance for those who suggested that as dominant a power as the U.S. needed to take a more subtle approach to its global affairs. Instead, beginning with the war in Afghanistan and continuing with its preparation and execution of the war in Iraq, America put the world on notice that global affairs would now be conducted on its terms alone.

America had made it clear that any threat to oil security is also a threat to its national security and that it will protect global oil supplies by any means necessary. In doing so, the Bush administration risks placing its armed forces in danger throughout the most treacherous and politically unstable areas of the world for decades to come. It is a scenario likely to be replayed not just in the Persian Gulf but also across the Central Asian republics, west Africa, Southeast Asia, Latin America, and anywhere else in the developing world where new oil is found. The net result may ensure America gets the oil it needs, but it is a policy that makes the United States secure, not safer. More U.S. troops will perish protecting oil and hatred of America will expand far beyond the ravings of Islamic extremists. Ultimately, it is an unsustainable policy.

6

Oil in the Family: The Bush Administration's Energy Policy

The golden age of oil is over regardless of how many billions of barrels still lie under the earth and deep below our oceans. For all the social progress oil has fostered, the technological innovation it has sparked and the economic wealth it has created, it has left a darker legacy—a trail of destitution, warfare, and environmental catastrophe.

The increasing global dependence on oil looks certain to bring about new civil conflicts. Even as our global economy suffers and the great nations of the world find themselves embroiled in a series of steadily escalating regional oil wars, another catastrophe looms in the form of global warming. Few scientists doubt that the steadily rising temperatures we have witnessed over the last century are caused by our increasing demand for and use of carbon-based fuels, most notably oil. No one yet knows what the full effects of global warming might be, but the prospect of violent new weather patterns and the melting of polar ice caps suggest a way of life few of us would wish upon our grandchildren.

If that isn't the future we want, then what can be done? The answer is both blindingly simple and infuriatingly complicated. The United States, along with the other oil-dependent nations of the world, must learn to conserve more while at the same time developing an efficient and self-sufficient alternative. No one is suggesting the U.S. abandon oil outright. Oil is just too important to every part of our daily lives. Indeed, given the very small amounts of oil necessary to drive the innovations of the petrochemical industry, oil could be a positive force in global society for decades to come. But overreliance on oil will be our ruin.

The United States has actually done a good job of conserving energy in some ways. Today the U.S. uses far less energy to run its power plants and heat its homes and offices than in the 1970s. In fact, energy consumption per dollar of the country's total gross domestic product is about half what it was in 1949. At the same time, the focus of the U.S. economy has moved to service and high-tech sectors. These need far less energy than traditional heavy industries. Yet, in one overarching way, the United States has undone all the progress it has made in other areas of energy conservation. American drivers consume more gasoline than any other people on earth and for the most part seem oblivious to the dangerous implications.

While gasoline is cheap there is unlikely to be a consumer groundswell to change government policy. Neither should we expect a road-to-Damascus conversion from the oil industry itself. Oil companies have spent so many decades operating under the roller-coaster mentality of boom and bust that all of the majors—even the so-called progressives BP and Shell—will keep pumping as much oil as they can, as quickly as they can, before they ever fully commit to a post-oil business plan.

Never before then has the United States, and the world, needed the strong leadership of a president who can resist partisan politics and the funding of special oil interests to move the United States beyond oil.

A successful renewable-energy strategy spearheaded by the United States would not only safeguard American jobs in the long term, it would lessen the chances of terror attacks on U.S. soil and ease the tensions that America's oil needs now place on so many unstable parts of the world. It could be a blueprint for the rest of the world, especially those new industrial powers like China and India whose oil needs are set to match and even eclipse those of the U.S. within the next few decades. It could also be a major step toward pulling the earth back from the catastrophe of global warming.

It is a bold and important challenge and it calls for a president who understands the importance of preserving the environment, one who wants to participate jointly with other nations in shaping our world and one who has the vision to embrace a world that flourishes without oil.

George W. Bush is not that president.

Bush's whole personal history has been shaped by the oil industry. His father, George Herbert Walker Bush, made his name in the Texas oil boom of the 1950s. The younger Bush grew up a product of the gung-ho wildcatting Texas oil culture.

He went on to launch his own fairly unsuccessful oil company before heading into politics. Bush staffed his administration with many boosters of the oil and automobile industry. And in nearly every energy decision he has taken in office—be it the push to drill for oil in the Arctic National Wildlife Refuge and other public lands, the EPA's wholesale gutting of the Clean Air Act in regard to power-plant emissions, the rejection of the 1997 Kyoto Protocol, or the ennui it has shown to improving fuel economy for cars and trucks—he has demonstrated that oil runs in his blood.

Oil in the Blood

The Bush family association with the oil industry began with Prescott Bush (George W. Bush's grandfather) who served as a director of one of Texas's most storied oil firms, Dresser Industries. In 1952, Bush's father, George Herbert Walker Bush, followed Prescott into the business when he founded his own company, Zapata Oil.

In 1963, Bush hired Houston attorney James Baker as his company lawyer shortly after Zapata merged with Penn Oil to form a powerful new independent concern, Pennzoil. The two men struck a fast friendship that would see Baker rise to the trusted position of secretary of state during Bush's single term as president. In 2000, the "Velvet Hammer," as Baker has been called for his smooth but ruthless style, came to the rescue of another Bush presidency when George W. asked Baker to be the public face of his campaign during the Florida recount.

George W. started another Texas oil company in 1978, Bush Exploration/Arbusto (Spanish for Bush). He had to be bailed out three times by his original investors before he sold Arbusto to another Texas company, Spectrum 7, in 1984. Bush remained as CEO of Spectrum but the new company continued to founder until it was snapped up in 1986 by Harken Energy. This time, Bush got to keep a seat on the board and received $600,000 in stock and a $120,000 salary from the deal. But Bush got his biggest break in 1988, the year his father became president, when Saudi investor Abdullah Taha Baksh took an 11 percent stake in Harken. Within months, Harken had won a contract to drill for oil in Bahrain despite having no international exploration experience. Few observers at the time doubted that Bahrain was playing a subtle game of politics. The contract, wrote the *Wall Street Journal* in 1991 "raises the question of . . . an effort to cozy up to a presidential son."

George W. Bush would later buy the Texas Rangers baseball team with his Harken windfall, giving him a high-profile position in Texas that proved to be the perfect stepping-stone for his successful 1992 run for governor of the Lone Star State.

Choosing the Right People

One look at the Bush cabinet shows that George W. Bush has stacked his deck with oil industry experience.

At first, Dick Cheney seemed a strange, almost desperate choice for running mate when Bush announced him in

August of 2000. At age fifty-nine, he was very much part of the old Republican guard, just as the party acknowledged it needed to appeal to a younger electoral base following Bob Dole's drubbing in the campaign of 1996. On top of that, Cheney had already suffered four major heart attacks.

But Cheney couldn't have had stronger credentials, as far as Bush was concerned. He was a conservative and he was passionate about the need to promote America's energy industries.

As Wyoming's congressman from 1978 to 1989, Cheney had cosponsored a measure to open the Arctic National Wildlife Refuge in Alaska to domestic oil drilling and he had voted against the Clean Water Act. After serving Bush's father as defense secretary during the 1991 Gulf War, Cheney quickly whipped through Washington's revolving door to became CEO of Halliburton, the oil-services company that made a fortune repairing Kuwait's damaged oil wells after the war. Cheney collected $36 million from Halliburton in 2000 while winning $1.5 billion in U.S. government loans for a company that in 2001 had a stock market value of $18.2 billion. But under his tenure, Halliburton also chalked up $23.8 million in oil rig and other equipment sales to the government of Saddam Hussein through European subsidiaries. After becoming vice president, Cheney sold his stake in Halliburton but his close ties to the firm caused a headache for the Bush administration when the Pentagon awarded a Halliburton subsidiary, Kellogg Brown and Root, a no-bid contract to rebuild Iraq's oil infrastructure.

Perhaps the only member of the Bush administration who

could rival Cheney's oil experience is National Security Adviser Condoleezza Rice. A brilliant political-science and foreign-policy wonk, Rice served Bush's father as director of Soviet and East European Affairs in the National Security Council and she spent six years as Stanford University provost. From 1991 until she joined the administration, Rice was also a board member of Chevron Corporation, during which time the company negotiated important Caspian drilling contracts with the former Soviet republics of Kazakhstan and Turkmenistan. In 1993, Rice had a 136,000-ton Chevron tanker named after her, following in the footsteps of another Republican bigwig, George Schultz. It was renamed the *Altair Voyager* in May 2001, after journalists started questioning Rice's association with a company that had recently been accused of human rights abuses in Nigeria.

For interior secretary, in charge of national parks and public lands, Bush chose Gale Norton, a lawyer who had a long history of representing industrial clients. Norton had worked for Delta Petroleum and had also lobbied for chemical giant NL Industries when it was fighting lawsuits over children's exposure to its lead paint. Norton began her career representing mining and oil companies at the Mountain States Legal Foundation. It was run by James Watt, later President Reagan's interior secretary and the man who famously downplayed environmental concerns by claiming there were more important matters to worry about, like the imminent second coming of Jesus Christ. Norton followed Watt to Interior, where she advocated opening the Arctic National Wildlife Refuge to oil drilling. Later, as Colorado's attorney general, Norton advanced a self-auditing procedure that allowed companies to police their own compliance with state and federal environmental laws. Norton was also the national chairwoman of the Coalition of Republican Environmental Advocates, a pro-business organization that was funded by Ford Motor Company and BP Amoco.

Commerce secretary went to Don Evans, an old friend of Bush from the oil industry and a longtime political confidante. Evans was campaign manager in both Bush's Texas gubernatorial races and his 2000 presidential campaign. Evans spent twenty-five years at Tom Brown Inc., a Denver oil company. Today, Tom Brown stands to benefit from new oil and gas development in the western Rockies.

Bush also chose two key cabinet members from the automobile industry. One was a longtime friend, Andrew Card, whom the president named White House chief of staff. Card was transportation secretary from 1992 to 1993 in Bush the Elder's administration. He also spent five years as CEO of the American Automobile Manufacturers Association, a lobbying group for the Chrysler Corporation, Ford Motor Company, and General Motors that, among other things, opposed stricter fuel emissions standards. In 1999, he became GM's chief lobbyist on Capitol Hill.

And for his energy secretary, Bush rescued Spencer Abraham after the single-term senator from Michigan had been defeated in his reelection bid. The son of an autoworker, Abraham was the top recipient of automotive industry campaign contributions in 2000, receiving $700,000 from General Motors, DaimlerChrysler, and Ford. As a senator, Abraham once tried to temporarily repeal the federal tax on gasoline to counter high summer fuel prices. He also cosponsored legislation that allowed consumers to deduct gasoline taxes from their annual income tax returns. Abraham was renowned as a friend of both the oil industry and the automobile industry and an advocate of more domestic oil drilling. Perhaps showing what he really thinks about federal oversight of energy issues, Abraham once cosponsored a bill calling for the elimination of the Energy Department.

Greasing the Wheels

The oil industry pumps millions of dollars into the coffers of politicians running for national and statewide office. Since the 1989–1990 election cycle, the oil industry (including the majors and independent producers and refiners, natural gas pipeline companies, gas stations, and fuel oil dealers) has given $115 million to political candidates and parties. Over the years, plenty of those recipients have been Democrats. But the vast majority of oil-industry money has gone to the Republican Party.

Since 1992, Republican candidates have consistently received twice as much hard and soft money oil-industry contributions as Democrats. In the 2000 presidential election, for example, George W. Bush received $1,929,956 in oil money while avowed friend of the earth Al Gore got just $138,014.

Spencer Abraham received $255,271 to go with his auto industry lucre.

In total, oil and gas companies gave $34 million to both parties during the 2000 election cycle. Topping the list of contributors were Enron, ExxonMobil, BP, Amoco, and Chevron, all of whom gave over $1 million apiece ($2.4 million in Enron's case) in a 75–25 percent split in favor of Republicans.

By 2002, the oil industry was still giving three-quarters of its money to the GOP but its total contributions had dropped off by over $10 million. Part of the reason was the lack of a presidential race, but it's difficult not to feel there might be an additional explanation: why would the oil industry need to contribute so much cash when its interests were already being so well looked after by the Bush administration?

Money at Work

The administration signaled its commitment to oil-industry concerns even before it officially took office. All but one of the sixty-three members of the Bush transition energy advisory team that was charged with helping formulate the administration's new energy policy came from some part of the energy industry. Prominent appointees included Enron Chairman Kenneth Lay, who had contributed $275,000 to the GOP during the presidential election campaign and who was still at this time a year away from his dramatic fall from power. Other industry groups represented included Hunt Power, Philips Petroleum Company, True Oil Company, and the American Petroleum Institute.

Within weeks of taking office, the president took world leaders by surprise, announcing that the United States would refuse to sign the Kyoto agreement on global warming. Bush said insufficient research had been conducted on the topic (a quorum of the world's leading climate researchers apparently wasn't enough) and that signing the pact would hurt the U.S. economy. Kyoto had been a thorn in the side of conservatives ever since it was adopted in 1997. For one thing, it stressed global cooperation—hence the chance for foreign nations to have a say in U.S. affairs—anathema to those who view organizations such as the United Nations as an affront to U.S. sovereignty. More important, it acknowledged global warming, which many conservatives continue to insist is just the product of a PR campaign trumped up by the liberal environmental movement to stymie big business.

The president did, however, signal his commitment to defining a long-term energy strategy by asking Dick Cheney to convene a task force that would develop a national energy policy.

The National Energy Policy Development Group, as the task force was titled, included all the senior members of the Bush cabinet. The group called major players in the electricity, oil, gas, nuclear, and coal industries for input while ignoring the views of environmental groups and other energy-industry watchdogs. In fact, of the top twenty-five energy industry donors to the Republican Party before the November 2000 election, eighteen companies met with the energy task force. The companies included Anadarko Petroleum, BP, Enron, the Exelon Corporation, FirstEnergy, and the Southern Company.

In May 2001, Dick Cheney unveiled the new National Energy Policy. The report rang the alarm bells for U.S. consumers. "America in the year 2001 faces the most serious energy shortage since the oil embargoes of the 1970s," the report said. If energy production increases at the same rate as it did during the last decade, then the country's projected energy needs would far outstrip its expected levels of production. And this trend, said the report, "if allowed to continue, will inevitably undermine our economy, our standard of living, and our national security."

The report considered the whole national-energy infrastructure including the use of coal, nuclear power, and natural gas as well as certain renewable forms of energy like solar power. But its assessment of America's oil needs was particularly sobering. By 2020, oil consumption would increase by 33 percent or 6 million barrels a day, while domestic production looks set to decline by 1.5 million barrels a day. So to meet projected demand, the U.S. would be forced to import 7.5 billion more barrels each day. The U.S. oil industry, once the powerhouse of world oil, would be able to supply just 30 percent of its country's oil needs.

To reverse this crisis, the Cheney report presented five national goals. It called on America to modernize the way it conserves energy, to modernize and overhaul the nation's energy infrastructure, to accelerate the protection of the environment, to increase national energy security by solidifying relationships with foreign energy producers outside of the Middle East, and to increase domestic energy supplies.

A crucial part of ensuring oil independence, the administration argued, was for the U.S. to start pumping more of its own oil. And the number-one prospect, it said, was Alaska's Arctic National Wildlife Refuge, a vast untouched wilderness area.

Alaskan Pipedreams

The Arctic National Wildlife Refuge (ANWR) covers some twenty million acres (about the size of South Carolina) and is the largest part of the National Wildlife Refuge System. It was established in 1960 to protect the land's "unique wildlife, wilderness and recreation values." All commercial activity, including oil exploration, was banned in the refuge. In 1980, President Jimmy Carter doubled the size of the range.

Only one and a half million acres of the refuge would be affected by oil drilling, but these happen to be on the northern coastal plain that is home to the richest concentration of biodiversity. Its ecosystem supports some 165 different species of birds, plants, and animals, including wolves, polar bears, and the 129,000 Porcupine River caribou who migrate there each year to give birth to their young. There are no roads in the refuge and no established campsites. It is, according to the U.S. Fish and Wildlife Service, "America's finest example of an intact, naturally functioning community of arctic/subarctic ecosystems."

There is undoubtedly a great deal of oil sitting under the refuge—some 20.7 billion barrels according to a 1998 U.S. Geological Survey study. But how much is economically and technologically recoverable is a matter of great dispute. A 2002 EIA report said that, in the most optimistic scenario, ANWR production would add as much as 1.5 million barrels per day by 2020 while in the worst-case scenario, ANWR would only add 590,000 barrels per day during the same period. The middle-ground estimates saw ANWR contributing 800,000 barrels a day to total U.S. production. Based on total projected U.S. oil needs, this would cut America's 2020 dependence on foreign oil imports by just two percentage points, from 62 to 60 percent.

According to the Sierra Club, ANWR holds just 3.2 billion barrels of recoverable oil when you take into account the economic costs of production. If the price of oil falls below $16 a barrel, then ANWR oil isn't even worth drilling. And based on current U.S. consumption, ANWR could satisfy just six months of the nation's oil thirst by itself. As the price of oil rises, of course, the more ANWR oil becomes economically viable. Even then, though, its production worth is likely to be diminished by increased global demand for oil.

Then there are the environmental concerns. No need to worry, insisted the administration. The oil industry's "footprint on the tundra" would be almost nonexistent, thanks to new "environmentally friendly exploration" techniques. These include horizontal drilling technology that minimizes the number of exploration wells companies need to drill and the use of ice roads and balloon tires to minimize the damage to the fragile ecosystem. The Bush administration also said it would put aside $1.2 billion in bonuses earned from ANWR oil leases to fund alternative and renewable energy research, and it would use royalties from drilling in the national wildlife refuge to fund other land conservation efforts.

Critics of the drilling were quick to point out that even though 3-D seismic research methods require less exploratory drilling, they still involve placing an enormous number of people on the ground in ecologically sensitive areas. Exploration calls for the presence of the huge thumper trucks that would be used to generate seismic measurements on the tundra. Also, even low-impact drilling requires a huge logistical operation to produce and transport the oil. Full production in ANWR could involve some 1,500 oil workers. They require housing near the refuge and transportation into the oil fields. In 2003, a report on Alaska's oil industry by the National Research Council found that oil companies had built 596 miles of road over the north Alaskan tundra and that seismic exploration had driven away bowhead whales, forcing the indigenous Inupiat population to venture further to hunt them for food. The report also noted the steep rise in obesity, alcohol abuse, and other health problems among the local population. And the report raised concerns that the oil companies would not adequately clean up the environment once they stopped drilling.

Despite a fierce lobbying effort by Republicans on Capitol Hill and the Bush administration, drilling in ANWR was not included in the 2003 Energy Bill. Political polls showed that some 60 percent of the U.S. public opposed ANWR drilling and while the GOP-dominated House was prepared to fly in the face of public opinion, Republican Senate leaders considered the issue a political pariah.

Go West, Young Man

The Cheney national energy report highlighted two main areas for exploration that past administrations had deemed unsuitable. The first included national parks, national monuments, and designated wilderness areas, all part of federal public lands that could hold some 4.1 billion barrels of oil and 167 trillion cubic feet of natural gas. The second included restricted parts of the outer continental shelf, the gently sloping undersea strip that separates the United States and the deep ocean, which is home to a possible 59 billion barrels of oil and 300 trillion cubic feet of natural gas.

America's public lands come under the jurisdiction of the Interior Department. During the Clinton administration, Western conservatives—notably ranching and antigovernment groups that are often referred to as part of the "wise use" movement—had vilified the department, led by Bruce Babbitt, even though the Clinton administration issued oil and gas leases to 23 million acres in public land: more than any previous administration.

By contrast Gale Norton has systematically given the green light to new oil and gas development on public protected lands throughout the Rocky Mountain states of Montana,

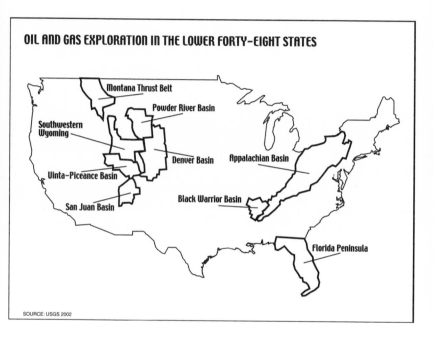

OIL AND GAS EXPLORATION IN THE LOWER FORTY-EIGHT STATES

Montana Thrust Belt

Powder River Basin

Southwestern Wyoming

Denver Basin

Appalachian Basin

Uinta–Piceance Basin

San Juan Basin

Black Warrior Basin

Florida Peninsula

SOURCE: USGS 2002

Colorado, Utah, New Mexico, and Wyoming, and in other resource-rich states like California and Texas.

In Utah, the Bureau of Land Management approved a request by the world's biggest seismic testing company, WesternGeco, to begin exploration work in the dome plateau region of the state on the eastern edge of Arches National Monument and next to Canyonlands National Park. The work called for thumper trucks to pound the desert soil looking for oil. On this same land, the bureau warns mountain bikers from cutting trails. The delicate sandstone spires found there make these Utah state treasures jewels of America's national park system.

Stunned environmental groups quickly mobilized to block the work. Even local companies were taken aback. "We're all just sort of shaking our heads because this area is so controversial," the president of Legacy Oil, owner of a drilling lease in nearby Lockhart Basin, told the *New York Times.*

Norton's department gave the go-ahead for fifty new drilling sites in areas such as Colorado's Canyon of the Ancients National Monument and San Juan-Paradox Basin, Wyoming's Powder River Basin, and even on Padre Island National Seashore, the nation's longest stretch of undeveloped beach and the principal nesting ground for the world's most endangered sea turtle.

Padre Island had seen oil and gas exploration in the past, but it had tailed off over the last two decades. The National Park Service had initiated a successful effort to save the Kemp's Ridley turtle from extinction. The Bush plan for new drilling allowed two new natural gas wells on the island, which required heavy trucks to drive back and forth over the turtle nesting grounds.

Interestingly, one of the only places where the Bush administration has refused to allow new oil exploration is Big Cypress National Preserve in Florida, where George W. Bush's brother Jeb is governor. In 2002, the Bush administration agreed to pay $235 million to buy back existing oil, natural gas, and mining leases in the area. It refused, however, to buy back similar coastal exploration leases in California.

In April 2003, the Department of the Interior took an even more radical step toward allowing greater oil and gas drilling on America's public lands. Norton decided to withdraw wilderness protection for three million acres of public land in Utah that oil and gas companies had long wanted to drill on. She also announced plans to limit Bureau of Land Management (BLM) wilderness areas to just twenty-three million acres. That was the amount of public land that the bureau evaluated back in 1991 when it completed a congressionally mandated review of 261 million acres of public land commissioned as part of the Federal Land Policy and Management Act. Since that time, millions more acres of public land had been earmarked for wilderness status, but Norton said the government could no longer consider protecting these lands.

In accordance with this new directive, the Department of the Interior suspended all ongoing wilderness protection evaluations of its Western land holdings, including large parts of Wyoming's Red Desert where natural gas companies hope to make a killing. And she revoked the Wilderness Inventory

Handbook, a policy guide that the BLM had used to assess wilderness land before it could be released for other uses. Essentially, Norton was washing her hands of the responsibility for creating any new wilderness lands.

The big question mark hanging over the Bush administration designs for America's public lands is the same one that bedevils attempts to open up ANWR for commercial oil and gas drilling—are the overall economic and national security benefits of this new domestic drilling worth the environmental risks to the nation's national parks and wilderness?

In the case of new oil exploration the answer is surely no. Even taking into account the new technology that can extend the life of America's once-mighty oil fields, there is little new oil left to be drilled onshore in the contiguous United States and the cost of producing it is so high that it represents a pretty poor investment.

Natural gas exploration, on the other hand, is more complicated. The environmental costs of producing natural gas are high, no matter what the industry and the Bush administration say. Wastewater runoff from natural gas production threatens to ruin grazing land that ranchers in Montana, New Mexico, and Wyoming depend on for their cattle.

But natural gas is likely to play an important role in America's next generation of energy use. Natural gas is the most cost-effective and environmentally safe way of powering the much-vaunted hydrogen fuel cells—the foundation for hydrogen-fueled cars. And the energy industry already relies heavily on natural gas to supply the nation's electricity grid.

Still, opening up America's wilderness lands is not going to help. A recent Energy Policy and Conservation Act study showed that 63 percent of all technically recoverable natural gas on federal lands lies outside national parks or wilderness areas. Only 12 percent of the reserves were unable to be leased under existing BLM restrictions.

As with ANWR, the Bush administration is spending a great deal of effort to tap what appears to be a very small amount of energy.

7

CAFE Society: The Brief Life and Untimely Death of Energy Conservation

Domestic drilling can't solve America's oil addiction. Consuming less oil will. Today, 97 percent of all the oil we consume is used for transportation and 70 percent of that goes toward keeping our automobiles on the road. Improve the efficiency of America's vehicles and you could make a good stab at breaking our oil habit.

Indeed, if we raised fuel economy standards to 40 miles per gallon for all our automobiles, including SUVs and light trucks, we could cut U.S. oil consumption by two million barrels a day over the next decade. That's one-tenth of the oil we currently use every day. Yet, passing new fuel economy legislation stands about as much chance as Congress voting to take a pay cut.

Twenty-five years ago, energy conservation was top of the list of concerns for U.S. industry. Since that time, the amount of energy America needs to produce $1 of gross domestic product (GDP) has been cut by nearly half. Companies like 3M have shaved 60 percent off their energy use while other firms have done everything from installing high-tech digital energy monitors to simply turning computers off at night. Increasing efficiency throughout U.S. industry means that less oil and natural gas are sucked up into producing electricity and America's total energy costs become leaner and meaner.

Somewhere in the middle of this conservation crusade, however, the automobile industry decided to turn back. And because of the auto industry's unequal share of U.S. oil consumption, our dependency on oil grew rather than shrank. In the process, the U.S. not only solidified its position as the number one global oil consumer but also established itself as the nation most responsible for greenhouse gas emissions. Not exactly the form of leadership the rest of the world was looking for.

It is pretty clear that American drivers crave big automobiles. They love the space to haul stuff around, they like the security and perceived safety they get from driving in large vehicles, and they like the idea of being able to transport Mom, Pop, the kids and grandparents comfortably in one car—even if it's rare that all members of the family actually want to be in the car at the same time.

But do big vehicles have to mean low fuel efficiency? No. All the major automakers could upgrade their vehicles without any need for further research and development. Certainly the public would support it—numerous surveys show that Americans welcome raising fuel-economy standards as a way of cutting dependence on foreign oil and reducing global warming.

Standing in the way is a formidable lobbying bloc consisting of automobile makers, car dealers, and the United Auto Workers of America. Together, they argue that improved fuel-economy standards will create smaller, more dangerous vehicles. New standards, they say, would undermine an already fragile automobile industry and cost thousands of jobs. And most of those jobs are located in Michigan, Ohio, and Illinois—three "swing" states that both Democrats and Republicans consider crucial to securing the presidency.

The Rise and Fall of Fuel Economy

In 1975, immediately following the Arab oil embargo, Congress passed the Energy Policy and Conservation Act. It established the corporate average fuel economy, or CAFE, standards for cars and light trucks sold in the United States. CAFE rules doubled the fuel economy of passenger cars from 1975 to 1985. Two years later light truck fuel-economy standards had also improved to 20.5 mph. But at that point, the makeup of the American automobile fleet began to change dramatically as demand for SUVs began to grow rapidly. It was spurred by a dramatic drop in the price of gasoline resulting from a worldwide effort to cut oil use and a production glut by the major oil-exporting nations.

Then in 1990, the U.S. suffered another oil shock as Saddam Hussein invaded Kuwait and oil prices shot up to nearly $40 a barrel. Capitol Hill was once again abuzz with talk about energy security and America's vulnerability to Middle East oil. Surely now was the time for America to come to its senses: the 1980s had lulled the country into a false sense of security when in reality, the U.S. was no more secure than during the dark days of the 1970s.

Richard Bryan, a Democratic senator from Nevada, certainly believed so and he introduced a bill to raise CAFE standards. But Bryan's bill was defeated and the story of how it failed shows how difficult it is to pass meaningful energy conservation laws in the U.S.

Bryan pretty much swung for the fences. He proposed a fixed 40 percent increase in each automobile maker's average car and light truck mileage to be implemented over ten years. That would have made Ford and GM boost their cars' mileage performance to 38.5 mpg by 2000 and their trucks' to 28.7 mpg. Foreign manufacturers like Toyota and Honda would have to produce cars that achieved 43 mpg because their fleets, with few light trucks, were already more fuel efficient.

It was a bold initiative, one, it could be argued, that spoke to the urgency of America's oil dependence. But, in reality, Bryan should have been thinking double rather than home run. Had he settled on a small increase of, say, 15 percent, there is a good chance that his bill would have passed. Light truck averages would have risen to 23.6 mpg by 2000 and automakers would have thought twice before releasing yet another under-performing SUV. Instead, Bryan's bill brought the auto industry roaring into life.

U.S. and Japanese carmakers began a major lobbying effort on Capitol Hill to block the bill. Detroit created a new advocacy organization, the Coalition for Vehicle Choice, and appointed a former Reagan administration official to head it. This "astroturf" group brought together more than 200 small business groups, all of whom swore that they needed large trucks to do their work and that the new standards would put them out of business.

Car dealers also protested the Bryan bill, claiming it would hurt their sales of big vehicles—their main source of profit. And the administration of Bush the Elder also opposed the bill, trotting out the old argument that fuel-economy improvements would lead to flimsier, smaller cars that put the U.S. driving public in danger. This despite the fact that, by this time, small cars were constructed with stronger alloy frames and effective crumple zones.

Still, the Bryan bill looked like it would pass. But then the Detroit automakers went into overdrive. They recruited the United Autoworkers of America to their cause, warning the union that the bill would mean tens of thousands of job cuts. As the bill came up for a vote on the Senate floor, the president of the UAW, Owen Bieber, flew to Washington, D.C. and personally lobbied senators against passing the bill. It was defeated by a single vote and with it disappeared the last real attempt to reduce America's dependence on oil.

The U.S. auto industry spent the rest of the 1990s aggressively developing the SUV market. SUVs are easily the most profitable part of the car business—Ford, for example, clears $20,000 profit per model in some cases—and it is the one area where U.S. automakers, thanks to import restrictions, have a clear advantage over foreign manufacturers. Other loopholes also drove the boom. The 1984 depreciation law, designed to crack down on small-business people who deducted their car as a business expense, exempted light trucks with a gross weight over 6,000 pounds. Owners of vehicles this large could deduct the full amount as it was assumed only farm equipment or heavy-duty work trucks would ever weigh that much. Then in 1990, Congress imposed a 10 percent luxury tax on cars costing more than $30,000. Again, thanks to auto industry lobbying, light trucks over 6,000 pounds were exempted. Finally, that same year, the industry persuaded Congress to treat light trucks more leniently than cars when it updated the Clean Air Act. The new law allowed many SUVs and light trucks to spew three times as much nitrous oxide as cars were allowed.

These tax breaks opened Detroit's eyes to a brand new way of making money—a new breed of large luxury SUVs that came with a built-in government rebate. Soon, the Ford Expedition, Lexus LX-450, and Lincoln Navigator were rolling off the production line. By 1996, some six out of every ten new luxury vehicles were SUVs. None of these supersized vehicles even came close to meeting the federally mandated fuel-economy standards for light trucks and most were hard pressed to do better than 14 mpg in the city. By 1999, light trucks, SUVs, and minivans made up 48.3 percent of new sales in the U.S. automobile market.

In May 2001, the Bush administration released its much-anticipated National Energy Policy specifically tailored to lessen America's dependence on foreign oil. The report spoke at great length about the need for new domestic oil production but it barely mentioned conservation. Instead, it said that Energy Secretary Spencer Abraham should base any new fuel-economy standards decisions on a new study by the National Academy of Sciences that was to be published later that year.

When it was released, the NAS report left little doubt about the effectiveness of fuel-economy standards. Current CAFE standards save the U.S. 2.8 million barrels of oil per day. It also concluded that, using existing technology, America's automobile industry could produce new vehicles that would increase fuel economy by eight to eleven miles per gallon within the next six to ten years without compromising passenger safety. Such improvements, said the report, would reduce America's dependence on oil, improve its terms of trade, and reduce the nation's sizeable contribution to global warming.

The NAS report also recommended putting the biggest SUVs on a crash diet. Slimming down SUVs and light trucks weighing over 4,000 pounds, it said, could improve overall highway safety by reducing the often-deadly mismatches caused in collisions between heavy and light vehicles. Yet despite the NAS findings, the Bush administration took only a minor interest in addressing fuel-economy standards. In 2002, it increased standards for light trucks by 1.5 miles per gallon. The changes take effect in 2004.

SUV RIP?

The attacks of September 11 and the subsequent war in Iraq sparked a renewed debate over America's dependence on foreign oil. The most obvious lightning rod in this debate was the SUV.

"Why the SUV Is All the Rage" was the cover of *Time* magazine's February 23, 2003 issue. The rest of the cover line—"They're family-size and fun, but gulp gas. Who's right in the war over America's favorite vehicles?"—highlighted the ideological battle being waged over the nation's highways.

The automakers chanted the mantra of giving the public what it wants, but they soon found themselves assailed from all sides. One week political pundit Arianna Huffington was launching the Detroit Project, a series of satirical anti-SUV ads she said was aimed at persuading "automakers to build cars that will get Americans to work in the morning without sending us to war in the afternoon—cars that will end our dependence on foreign oil." Another week, southern Baptist minister Jim Ball was preaching an anti-SUV message titled "What Would Jesus Drive?"

The Sierra Club started ridiculing GM's monster SUV, the Hummer H2, in an animated online campaign while Rainforest Action Network launched a full-court press against Ford Motor Company, accusing it of having the worst global-warming polluting cars. Back in 2000, Ford had pledged to unilaterally raise its fuel-economy standards by 25 percent by 2005. Stunned by Ford's initiative, both DaimlerChrysler and GM said they would match what Ford did. But in February of 2003, Ford acknowledged it couldn't meet its own goals.

In East Coast cities, environmental activists were plastering SUV windshields with mock traffic tickets that read DRIVING

AN INTERNAL COMBUSTION AUTO IS HAZARDOUS TO YOUR HEALTH & THE HEALTH OF THIS PLANET. And in Los Angeles, a few people took their dislike of SUVs to the extreme. In the summer of 2003, over seventy new Hummer H2s were set aflame and destroyed at car dealerships around the city. A group calling itself the Earth Liberation Front claimed responsibility for the attacks.

Few in America, whatever side of the fuel-economy debate they might sit on, agree with the torching of SUVs. But there was clear evidence that Americans did care about their addiction to imported oil. One 2002 poll of likely Michigan voters found that 74 percent favored stronger fuel-economy standards and that support crossed party lines. Of households that included United Auto Workers, nine out of ten polled favored stronger regulations. Another national poll that year reported that 88 percent of likely U.S. voters supported increasing fuel standards for cars and trucks. Yet another poll reported that more than 68 percent of likely voters believed energy efficiency should be a higher priority than energy production. That was an eighteen-point increase in support for fuel efficiency since a similar poll was conducted before the attacks of September 11.

The statistics were heartening for those advocating stronger fuel-economy standards but they were also depressing—many of those people who supported better fuel standards were also buying SUVs. In 2002, SUV sales rose 6.9 percent to nearly four million, even as total passenger vehicle sales fell 1.9 percent. Meanwhile the collective fuel economy for all U.S. automobiles dropped to 24 mpg, its lowest level since 1980. American drivers wanted it both ways. They cared about lessening their dependence on foreign oil but they still had to have their SUV.

In late 2003, the Bush administration finally announced its intention to redress CAFE standards. In an effort to boost fuel economy and improve road safety, the administration suggested closing some of the SUV loopholes, notably removing "crossover" SUVs (those built using car bodies rather than light trucks) from the light truck category. It also proposed lifting the exemption that had allowed vehicles weighing over 8,500 pounds (the H2, for example) to avoid any fuel-economy regulations.

The most sweeping idea, however, was a plan to classify new light truck fuel-economy standards by weight. The lighter the truck, the higher the gas mileage it would have to attain. Parsing the administration's thinking, it seemed the government was floating some compromise measure that would look good to the public while also keeping the automobile industry, if not happy, at least mollified. Making light trucks and SUVs match the fuel-economy levels of cars would never get past the auto industry lobby. This way, Detroit's Big Three would have to make some changes, but they would still be able to keep building their largest (and most profitable) trucks. And the administration could still claim it was tackling America's oil dependency.

Environmental groups, led by the Sierra Club, cried foul. By creating more truck weight classes, it argued, automakers would have even more incentives to build heavier, less fuel-efficient vehicles. "The loophole the Bush administration proposes to close may move us one step forward, but the one they propose to open for the auto industry will likely move us three steps back," said Daniel Becker, director of Sierra Club's Global Warming and Energy Program.

It wasn't just the environmentalists who were angry. The Autoworkers Union of America warned that the new rules would cause massive job cuts as Detroit would no longer have any incentive to manufacture the small cars that helped it meet average fleet fuel standards in the past.

A Hybrid Solution?

If the U.S. is going to make any inroads into cutting its dependency on foreign oil before those demands completely dominate its economic and foreign policy agenda, it will need to embrace a new generation of hybrid-powered automobiles.

Hybrids rely on a combination of gas engine and electric motors to increase fuel economy anywhere from 15 percent to 60 percent above the performance of a gasoline engine. In general, the electric motor powers the car at low speeds up to 20 miles per hour and then the gas engine kicks in to provide more power. Likewise, the gas engine will add power on inclines or whenever extra power is needed. Hybrids use what is called a regenerative braking system. The electric battery recharges as the car slows down.

In 2003, only two companies, Toyota and Honda, offered hybrid vehicles in the U.S. Together, they have sold more than 155,000 of Toyota's Prius and Honda's Insight and Civic hybrid worldwide since 1999, but that's still a tiny number when you consider that seventeen million new cars are sold each year in the U.S.

Part of the reason so few hybrids have been sold is the cost. On average, a new hybrid costs around $3,000 more than a conventional gasoline-powered vehicle. That extra cost is mainly due to the low economies of scale that come from producing a small number of vehicles.

Hybrid costs could be reduced if they were to become more popular and consumers could be given an incentive to buy these new cars with the help of a tax break. So far, the U.S. government offers a $2,000 tax deduction on each hybrid, a paltry amount when you consider that the government allows owners of luxury SUVs to write off the full cost of their vehicle as a business expense.

Indeed, a budding demand for hybrids already seems to be in place. According to JD Power and Associates, the leading market-research firm for the automobile industry, 30 percent of 5,000 people they surveyed in 2002 said they would definitely consider buying a hybrid.

Having a hybrid already has its benefits. Hybrid owners in some cities are allowed to use rush hour express lanes reserved for car pools no matter how many passengers are in the car. And hybrids (the Prius in particular) have captured the imagination of the Hollywood crowd—the same opinion and style makers that helped make SUVs seem so trendy just a few years ago.

Supporters of the technology are confident that hybrid cars and SUVs will soon get over 60 miles per gallon and even U.S. automakers like Ford believe hybrids will provide a crucial stepping-stone from reliance on pure gasoline engines to a new form of energy such as hydrogen fuel cells. In 2004, Ford unveiled a hybrid version of the Escape, its small SUV. Toyota has even bigger dreams—it hopes to release hybrid models of half of its existing car and SUV fleet.

Getting America to switch to hybrid technology will take more than better fuel economy and cleaner air, though. Successful hybrids will also need improved power. Because hybrids depend on electric motors to move from a standing start, they have good low-end torque—or initial acceleration—from 0 to 30 mph. Electric motors produce their best acceleration at low speeds and have none of that clunky gear shifting that comes with a gasoline-only engine. Hybrid engines also create very few vibrations and there is less need for brake and engine maintenance than in gasoline engines because regenerative braking technology creates less wear and tear on brake pads and discs.

But hybrids won't work for everyone. Most notable among those likely to be underwhelmed by hybrid technology are drivers of big SUVs and light trucks—exactly the people who will need to be wooed from gasoline if America is going to beat its oil habit. Hybrid technology delivers less power at high speeds, on steep hills, and when called upon to tow a heavy object—say a boat or trailer. When it comes to power, hybrids might have the jump on an H2 at the traffic lights but they are no match for a Dodge with a Hemi.

Which might explain why automakers continue to flirt with diesel engines, an updated old technology that promises great fuel economy and plenty of muscle. In Europe, new clean diesel technology already powers one-third of all new cars sold. By 2006 that figure could rise to 50 percent.

Diesel has definite advantages over gasoline. Diesel engines get better torque than gasoline equivalents and much better fuel economy—up to 50 mpg on highways—which helps explain their popularity in Europe where gas prices can run five times as high as those in the U.S. They also contribute less global warming carbon dioxide emissions—30 percent less in some cases—than gasoline engines. But diesel engines are more expensive to produce than gasoline ones—it's those pesky economies of scale again. They also pump higher levels of nitrous oxide—a main precursor to smog—into the air than gasoline engines as well as other pollutants that cause asthma and respiratory problems.

New emissions standards passed by the state of California, along with federal tailpipe standards adopted in 2002, will require large cuts in diesel engine emissions. This means major diesel manufacturers like VW won't be able to sell many of their vehicles in the U.S. until they conform to the standards. Despite these drawbacks, "clean diesel" engines look set to play a stopgap role for U.S. automakers until a true clean, fuel-efficient, and high-performance automobile comes along.

You Pay How Much for a Gallon?

Even if hybrid technology does succeed in capturing the imagination of the American driving public, it could take a decade before it helps cut U.S. oil consumption. Even then, the benefits would only apply to new cars coming onto the market. That still leaves a generation of durable secondhand cars on the road. Any improvement in fuel economy gained from hybrids and tougher CAFE standards would still be undercut by these aging gas guzzlers.

However, there is a way of making us reduce the amount of gasoline we currently use. The strategy is proven and it would produce results within hours of being implemented. What's more, it is an idea endorsed by a host of American experts. There's just one problem. It involves raising taxes on gasoline.

The average American driver buys 690 gallons of gasoline a year and in 2000, U.S. drivers purchased a total of 130 billion gallons. Part of the reason American motorists drive so much is that the cost of gasoline is so reasonable. Even as gas prices hit nearly $3 a gallon in some parts of California over Labor Day 2003, they were still at least 50 percent cheaper than the price European drivers pay at the pump.

The difference in price isn't due to transportation costs—most European oil is shipped from the Middle East or North Sea while the U.S. increasingly relies on suppliers like Mexico and Canada. And it isn't that Europeans pay more for crude oil than the U.S.—they don't. The main reason is taxation—Europe has been paying high taxes on gasoline purchases since the early part of the twentieth century. Back then, cars were considered a luxury item in Europe, and slapping a hefty tax on gasoline was not considered prohibitively expensive because Europeans didn't drive the same distances as Americans. European governments were acutely aware that they didn't have their own supplies of oil so they used taxation to control demand.

Europe consumes nearly six million fewer barrels of oil a day than the U.S. So wouldn't it be worth a shot to raise federal gasoline taxes to European levels? If nothing else, higher taxes would make sure that luxury SUVs really were a luxury item. It would cost over $150 to fill them up, after all. Politically this might seem unfathomable, but a growing number of people inside and outside the energy industry say it might be the only short-term way to wean the U.S. off its oil addiction.

At the moment, the price we pay for gas fails to take into account any of the many indirect costs associated with oil and automobiles. These include the huge subsidies and tax breaks that government doles out to the oil industry each year as well as the social, economic, and environmental costs of keeping America's cars on the road. Incorporate these indirect costs into the consumer price of gas and a 100 percent tax hike starts to look palatable by comparison.

Placing a price on social and environmental factors is tricky at the best of times. For one thing, why should we count oil's external influences as a negative factor? In its short history, oil has played an amazing role in improving the quality of life in the U.S. and much of the rest of the world. Without oil, we would be deprived of many of the amenities we now consider crucial to modern life.

The flip side, of course, is that oil comes at a very high cost. It takes a toll on the environment and air quality. Securing large supplies of oil around the world involves billions of dollars in military and diplomatic costs. And oil exploration comes at a huge social cost to thousands of communities all over the world. And that's before you take into account the tax breaks and subsidies the U.S. government gives oil companies each year.

In 1998, the International Center for Technology Assessment, a scientific and environmental think-tank that included experts from the Union of Concerned Scientists and the Sierra Club, attempted to quantify the many costs of using automobiles that are not reflected in the price Americans pay for gas at the pump.

Using 1997 statistics, it reported, the federal government handed out tax breaks and subsidies in excess of $10 billion to the oil industry to cover such things as domestic oil-field depletion, the production of nonconventional fuels (or heavy oils), exploration, development costs, and foreign tax credits.

Federal, state, and local government also provide annual subsidies for petroleum extraction and use that can run anywhere from $438 million to $114.6 billion. These include the vast sums spent on transportation infrastructure including the construction, maintenance, and repair of roads and bridges. Other subsidies support oil industry research and development and the role played by the Army Corps of Engineers and the Department of the Interior's Oil Resources Management Programs. Then there are other government expenditures such as cleaning up oil pollution and other oil-related liability costs.

Subsidies related to protecting America's oil interests also add up. In the summer of 2003, the U.S. was spending $1 billion each week on military operations in Iraq. When the administration finally asked Congress to finance a one-year reconstruction project for Iraq and Afghanistan, the cost was a whopping $87 billion, with the majority designated for military operations. Add that to the costs of the first Gulf War, and of maintaining U.S. Central Command (the branch of the military responsible for Persian Gulf security), and the cost of protecting our global interests looks very high indeed.

Environmental, health, and social costs, the center said, run anywhere from $231.7 billion to $942.9 billion each year. That included the effect of automobile pollution on human health (back in 1993, the Worldwatch Institute estimated the damage to human and environmental health from vehicle emissions was $93 billion a year), as well as noise pollution and the ancilliary effects of urban sprawl. A separate report estimates that urban sprawl adds an average of six pounds in weight to each person in America. The resulting health costs that come from heart disease and other weight-related illnesses would also be included in the real cost of gasoline. Then there's global warming. Some 4 percent of world greenhouse gas emissions come from U.S. automobiles; one insurance company study projects that climate change will cost the global economy an additional $304 billion in direct costs every year.

Finally, the center considered other costs associated with road travel. These included delays due to congestion, uncompensated damages from car accidents, as well as the cost to the U.S. economy caused by sudden swings in the price of foreign oil. All told, the center estimated that these external costs could total anywhere from $558.7 billion to $1.69 trillion. That would add between $4.60 and $14.14 to the retail price of a gallon of gasoline.

Now, should we take into account all of these external costs when evaluating the real price of gas? Not necessarily, but these examples provide a good amplification of what America's oil dependence means in real-world terms. And the International Center for Technology Assessment is hardly alone in its evaluations. One California State University economist told *Business Week* that the real cost of oil is at least $13 a barrel more than it currently costs. That would take crude oil prices up to $45 a barrel and prices at the pump near $4.50 a gallon. Even then, the Cal State calculations didn't even take into account the costs of air pollution, oil spills, or global warming.

8

Beyond Oil: The Hydrogen Balm

In June 2003, the Ford Motor Company threw itself a blowout party. The company founded by Henry Ford in 1903 was intent on celebrating its centennial in style. So Ford transformed its 152-acre Dearborn, Michigan headquarters into a parade ground of old and new Ford vehicles, including 100 original Model T's that had set off in convoy from California the week before. Other attractions included a night of NASCAR racing and a series of concerts starring Beyoncé Knowles as well as the official Ford pitchman, country music star Toby Keith.

For the hundreds of press who had come from all over the world to be part of the Ford festivities, the company had arranged a sneak preview of the new Ford Rouge Center, a twenty-first-century reinvention of Ford's legendary assembly plant.

The original Rouge factory opened in 1917 and was named after the river that ran through it. The Rouge was a marvel of modern technology, and heads of state, movie stars, and tourists by the carload came to see the vanguard of the mass-production factory. Ford forged steel on the premises and every stage of vehicle assembly took place under its roof. What went in one end of the Rouge as iron ore, sand, and water emerged from the other as a fully assembled Model A.

For more than seventy years, the Rouge churned out vehicles like Matchbox toys. The new Rouge, centered around the Dearborn Truck Plant, will also be a pioneering piece of engineering. Thousands of F-150s, America's bestselling trucks, will rumble out of a hangar with the world's largest living roof—a 10.4–acre sylvan carpet of sedum vegetation. The green rooftop absorbs rain and snow, reducing storm-water runoff into nearby rivers while trapping carbon dioxide to keep air around the plant relatively pure.

As I sat listening to the Ford executives extol their new factory as part of the company's "green expression to the outside world," I couldn't help but be a little impressed that they did so with a straight face.

After all, here was the birthplace of one of the most polluting and fuel-inefficient trucks in America, and Ford was claiming an environmental victory all because it had built a "green" roof.

Going Green

In today's business world, most big industrial companies realize they need to develop an environmentally friendly public face. George W. Bush may not believe that global warming exists, but, thanks to an energized and increasingly media-savvy environmental movement—not to mention a wealth of scientific data—a good part of the United States does. And corporate America, if nothing else, is very attuned to the sensibilities of its customers.

Simply cast a glance at the Web sites, print, and TV advertising campaigns of all the major oil companies and automakers and you'd be forgiven for thinking you'd stumbled on promotional material for some new nature refuge. Bunny rabbits and eagles live in perfect harmony with intrepid oil-company workers, all of whom look more like they are modeling an adventure gear shoot for *Outside* magazine than conducting the very grubby business of drilling for oil. SUV ads show how you too can commune with nature—with the help of all-wheel drive and a hefty V-8.

PR aside, Shell is part of a groundbreaking project in Iceland to establish the first self-sufficient hydrogen fuel economy. BP, meanwhile, took steps to reinvent itself back in 1997 when chairman Lord John Browne shocked the rest of the industry by withdrawing from a group that denied global warming.

Since then, BP has moved energetically to recast itself as an energy company rather than an oil company. It controls 17 percent of the world's solar energy market and offers photovoltaic (solar panel) services in over 160 countries. It has invested $10 million to create a photovoltaic power system for 1,852 schools in Brazil and $3 million to power a solar telecommunication system in remote areas of Peru. The company has also put money in wind power and hydrogen. But perhaps its most effective strategy has been to undertake a dramatic rebranding of its corporate image. BP spent over $600 million on its Beyond Petroleum marketing campaign to emphasize its green credentials.

Critics of the oil industry dismiss these green strategies as bluster, a cynical attempt by a cynical industry to plaster over the decades of environmental and human rights damage it has caused. Environmentalists point out that BP was, for many years, the leading force behind opening the Arctic National Wildlife Refuge to oil exploration (it withdrew support in 2002), and recent Siberian deals with Russian companies have landed the company in the middle of one of the biggest oil-induced environmental disasters in the world. Many outside observers have come to question just how committed BP is to renewable energy. Even Lord Browne is more bashful than he was before. BP, he recently told a Sierra Club roundtable, "stands for BP." The phrase Beyond Petroleum was not mentioned.

Nearly 150 years after the first oil well was drilled, there is an urgent need for global society to move beyond oil. We need to switch to a renewable form of energy—one that doesn't leave the world's great economies dependent on the Middle East, one that doesn't foster civil war, inequality and poverty, and one that doesn't threaten the health of our planet.

The oil companies aren't stupid. They understand better than most that the window of finding new streams of oil is closing, and they also know that the consumers in the developed world genuinely believe that oil dependence is damag-

ing the environment and endangering national and global security. Most important, if Big Oil doesn't get involved now
with renewable energy, it might be shut out later on. The
history of oil has shown that global energy tends to be dominated by a few multinational powers with the reach and financial clout to supply energy all over the world. No one can
foretell what industry structure oil's successor will dictate, but
you can be sure that today's oil giants have every intention of
holding on to their share of the energy market.

Still, for even the most progressive supermajors to shift to a
renewable strategy would take years. The oil companies are
about as agile as one of their supertankers. And the captains
of these corporate behemoths still seem unsure that now is
the time to make the turn.

The reason is simple. Oil has been written off countless times
during its short but eventful life. Every few decades bring with
them a new scare that oil is running out and prices skyrocket
only for new rich streams to be discovered accompanied by a
price glut. Oil has always followed this boom-and-bust cycle
and for the most part oil companies have learned to remain
sanguine about the future while pumping as much now as
they can. No one really believes that huge new fields like
those in Saudi Arabia or Iraq will ever be discovered again but
the oil companies are confident that there is still a great deal
of money to be made from oil—far more at the moment than
could be made by switching to renewable energy.

Even if the oil-depletion soothsayers are correct, Big Oil
still has little incentive to move beyond petroleum. The peak

oil argument presents a doomsday scenario where oil prices continually rise and where, over time, geopolitical power and market dominance shift permanently back to the Middle East. That's terrible news for U.S. consumers and the government. It bodes ill for the economy and it increases the chances that U.S. troops will become embroiled in new oil-related foreign conflicts.

But when oil prices rise, oil companies make a lot of money. Even if conventional sources of oil get priced out of the market, oil companies can still turn to the so-called unconventional or heavy oils—like deep-sea oil, oil shale, and tar sand. The high price of oil also makes the development of synthetic oil from coal deposits attractive. In the short term, at least, higher oil prices might actually increase the amount of oil available to the global market.

Good news, right? No, not really. Mass production of heavy oils like tar sand may delay OPEC's stranglehold on the global economy but it will take a far more devastating toll on the environment than conventional oil production.

Tar sand production requires an enormous amount of water, the other natural resource that looks set to dictate our future. Often, the sand has to be heated to extract the oil. That means extra energy has to be used just to produce oil. This extraction process creates waste sludge that cannot be reclaimed and contains hydrocarbons, inorganic salts, and heavy metals. By the year 2020, Canada's major tar sand producers will have accumulated one billion cubic meters of this waste tailings, as it is called. Currently, the waste is stored in protective ponds but they remain toxic for over 100 years and were they to leach into groundwater or the soil they would have a devastating impact on the environment.

Heavy oils like tar sand and synthetic oil extracted from coal also pose a global problem. Producing usable oil from heavy oils creates large amounts of carbon-dioxide emissions and so increases the greenhouse gases that cause global warming. If heavyweight oil is the great black hope of the oil industry and the world's demand for energy doubles by 2040 and triples by 2070 as many projections say it will, then our oceans, climate, and atmosphere may well suffer total knockout years earlier than even the most doomsday pundits have yet forecast.

Hydrogen to the Rescue?

"Tonight I am proposing one-point-two billion dollars in research funding so that America can lead the world in developing clean, hydrogen-powered automobiles."

That was how President Bush announced his commitment to move the United States beyond oil in his State of the Union speech in January 2003.

Such a categoric endorsement of a technology that is so threatening to the interests of Big Oil took many of the president's critics completely by surprise. Maybe, finally, the president was showing real leadership in helping to break America's addiction to oil.

The Bush initiative was presented as part of a collaboration with the auto industry to build a hydrogen fuel cell vehicle dubbed FreedomCAR. The project aimed, so said a promotional Web site, to give the U.S. "an historic opportunity to develop technologies that could lead to a personal transportation system that uses renewable energy resources. Success will establish the United States as a global leader in environmental and energy technologies and will be a key to ensuring future U.S. competitiveness."

Support for hydrogen cuts across political boundaries. The automakers say they are investing heavily in it, environmentalists have been touting it for years, and even the oil companies seem to think it's a smart idea. Which makes hydrogen fuel cells a no-lose issue for a president whose green credentials are virtually nonexistent. But if the prospect that these historically antagonistic groups will now stop battling and agree to share a ride on hydrogen seems too good to be true . . . it is. When you take a closer look at hydrogen, it becomes apparent very quickly that the oil and automobile industries on the one hand and the environmentalists on the other are not talking about the same thing at all. You can figure out for yourself where the Bush administration stands on the issue.

The idea of using hydrogen—the most basic and plentiful element in the universe—as a way of producing fuel came from an eccentric nineteenth-century Englishman named William Grove. While experimenting with a new process called electrolysis that separated water into its core elements of hydrogen and oxygen, Grove hit on the idea of doing just the reverse. In 1839, he combined oxygen and hydrogen across a pair of platinum electrodes and produced an electrical current. The only by-product was pure and drinkable water. Later he linked together a number of these devices and came up with a rudimentary fuel cell that could store enough hydrogen to produce electric power.

Grove didn't build on his breakthrough though science fiction master Jules Verne imagined a world powered by hydrogen in his 1874 novel *The Mysterious Island*. But hydrogen remained very much the stuff of Jules Verne until the 1960s when General Electric began playing with Grove's ideas. The fuel cell as we know it today was born when GE developed monster-sized cells as power supplies for the Apollo and Gemini space missions. Those cells also produced fresh drinking water for the astronauts. NASA still uses fuel cells on its space shuttle missions, and many office buildings rely on fuel cells at least partly for electrical power. But it is hydrogen's potential as a clean and endless fuel supply for automobiles that has captured so many imaginations.

GM, along with the other members of the Big Three—Ford and DaimlerChrysler—are all part of the FreedomCAR project. GM already has a prototype minivan called the HydroGen 3 that is the heart of a public relations campaign. The HydroGen 3's platform base, a vacuum-insulated stainless-steel canister that houses fuel cells and the supercooled liquid hydrogen that feeds them, could conceivably be used for all of GM's cars. Much like the original flatbed SUV construction,

GM envisions mounting a variety of different car and light truck bodies on the same fuel-cell platform. The ride, however, will be a lot smoother.

The fuel cell powers the vehicle like this: the electrochemical reactions begin when hydrogen enters one side of the fuel cell. There, the hydrogen is separated into electrons and hydrogen ions. In the case of one popular type of fuel cell—the proton-exchange membrane used by NASA back in the 1960s—the ions pass through a membrane to combine with oxygen on the other side, making water in the process. The electrons, however, can't pass through the membrane and so are forced through an electric motor. As they pass through the motor, the electrons transfer power from the fuel cell to the motor and the motor drives the wheels of the car.

A hydrogen-powered car, because it relies on electrical not mechanical power, is much smoother to drive than a gasoline-powered internal combustion engine. It also propels the car nearly three times as far using the same amount of energy. And because there are very few moving parts—the fuel cell will be hooked up to a computer processor that works the lights, drive-train, braking functions, and so on—running repairs should be kept at a minimum. Of course, with a computer running things, motorists will now have a second type of car crash to worry about.

Environmentally speaking, a fuel-cell car would produce no harmful emissions, just a steady trickle of clean water. Geopolitically, fuel cells could be the breakthrough in winning independence from Middle East oil.

The key word here is *could* because, despite all the booster-ism coming from the auto industry (GM says it has spent nearly a billion dollars getting its fuel-cell vehicles ready for the road and would like to sell a million each year by the middle of the next decade), the hurdles hydrogen must overcome before it topples oil from its seat of power are immense.

One is how to store enough hydrogen in the fuel tank to match the miles that cars get today on a full tank of gas. The easiest way to store hydrogen is as gas, but that takes up a lot of room. So the gas must be compressed to fit into a car tank. That calls for a tank that can withstand as much as 20,000 pounds per square inch of pressure to prevent it from busting in a crash. Liquid hydrogen poses its own problems. It puts less pressure on the tank but it needs to be cooled to –423 degrees Fahrenheit. You need a great deal of insulation, which increases the size of the tank. Even then, a hefty "angels' share" of liq-uid evaporates daily, leading to a buildup of pressure that has to be bled off. Liquid gas fuel-cell prototypes that aren't topped up every night can "run out of gas" without leaving the garage.

Another option might be to fill fuel-cell tanks with a solid material like lithium hydride or sodium borohydride that can soak up hydrogen like a sponge and release it as needed. These solid materials take up far less space than gaseous hy-drogen and can be stored at room temperature. But energy is required to infuse the material with hydrogen and solid sub-stances often only release the hydrogen when heated to very high temperatures. That hampers the car's performance and who wants a next-generation Pinto?

Then there are the safety issues. Fuel-cell drivers are unlikely to suffer the fate of those aboard the Hindenburg, the hydrogen-filled German dirigible that went down in flames over New Jersey in 1937. But hydrogen is very flammable and burns without color or odor. The earliest gasoline-fueling stations would pour the gasoline through a gauze cloth to keep the vapors from igniting. At present, when Daimler-Chrysler employees refill the company's prototype Mercedes-Benz fuel-cell car, they first put on anti-static lab coats to minimize the chance of a spark. The company also uses a digital communication cable to measure the pressure and temperature of the hydrogen gas being dispensed.

Aside from all the technical and safety issues that have been solved, there is still the conundrum of how we build a hydrogen fuel infrastructure to satisfy America's motorists.

Hydrogen-producing units can be housed anywhere. Conceivably, fuel-cell car owners could make hydrogen at home or fill up at community refilling sites. But, at first glance, it would appear that the oil companies are best equipped to sell hydrogen fuel. In recent years, all the majors have put more emphasis on building gas stations and marketing their products. BP and Shell have also invested hundreds of millions of dollars in hydrogen production and storage, while other oil companies are extracting hydrogen from gasoline at nine refineries around the nation for industrial purposes. For the consumer, filling up their new vehicles at gas stations they already frequent could prove crucial to making hydrogen cars a success.

But hydrogen providers, be they major energy companies like Shell and Exxon, or new independents, are not going to build those hydrogen filling stations unless there are enough fuel-cell vehicles on the road. And automakers aren't going to build a new fleet of fuel-cell vehicles unless there is a national network of hydrogen-fueling stations.

Getting that sort of new energy grid off the ground is too big a task for even the combined might of the oil and automobile industries. But the Bush administration has made no provision for this in the FreedomCAR initiative. Why should the federal government subsidize the start of a fueling infrastructure? It's a valid question. After all, gasoline-powered cars didn't need any such push in the early twentieth century.

But that was when there were no automobiles. Such was the novelty of the horseless carriage that all of America got caught up in car culture. The allure of independent travel created its own demand. Yet even then, cars received a helping hand. FDR's New Deal invested millions in roads, and Eisenhower's National Defense and Safety Highway Act created an Interstate Highway system that cemented the role of the automobile in American life.

Which makes the transition to a new type of fuel even more difficult today. Gasoline is cheap. Most conventional automobiles will be more affordable than any new fuel-cell vehicle that first comes on the market. As the hybrid experience has shown, manufacturing economies of scale dictate that these new vehicles will be expensive to produce until demand grows and production costs drop. It's a classic Catch-22, but the federal government could help matters by taking back some of the millions of dollars in oil industry tax breaks. We could use the money to build demand for fuel-cell vehicles.

The government could also encourage the military, metropolitan transportation systems, and emergency services to fast-track hydrogen fuel cells, especially where they can easily be serviced by a hydrogen refinery hub. A glimpse of what could be America's energy future can be glimpsed in Iceland. Icelandic New Energy—a consortium formed in 1999 and including DaimlerChrysler, Shell Hydrogen, and Norsk Hydro as partners—recently launched the world's first commercial hydrogen transportation project. It aims to run three Daimler-Chrysler hydrogen-powered buses on the streets of Reykjavík, all depending purely on a new Shell hydrogen station for fuel.

Shell and DaimlerChrysler's collaboration in Iceland produced a working hydrogen vehicle and infrastructure in just four years, albeit on a small, experimental level. The Reykjavík prototype is set to run until 2005 and cost about six million dollars—two million coming from an EU subsidy and the rest fronted by the participating companies.

The FreedomCAR project, on the other hand, has no plans to build a commercially viable hydrogen car before 2015. Even then the project won't address the problems of establishing a hydrogen-refueling infrastructure. And though the Bush administration has pledged $1.7 billion in federal funding, critics of the government and the auto and oil industries complain that FreedomCAR may just be a ruse to stall any meaningful steps beyond oil. They point to the auto industry and government's intransigence on improving fuel economy for cars being manufactured right now and the fact that Detroit has promised a miracle-mobile before—and failed to deliver.

In the 1990s, the Big Three joined with the Clinton administration in forming the Partnership for a New Generation of Vehicles. It was supposed to produce a family sedan capable of getting eighty miles to the gallon and the federal government spent $1.5 billion on the project. By the time it was scrapped, just three prototypes had been built.

"I'll throw out a very cynical perspective which happens to match what we have seen from General Motors over the past several years," National Resource Defense Council analyst Ronald Hwang told the *New Yorker* in August 2003. "They're trying to bedazzle us with very fancy prototypes to get us hooked on the fact that they're working very hard on a technology that we all agree is ultimately the prize—a zero-pollution motor vehicle. At the same time, General Motors is on record saying, 'Please don't make us improve the fuel economy of our current gasoline fleet.' They're trying to undermine political momentum to do something that is going to have a real impact over the next decade or two."

Hwang's point is well made. A commercial fuel-cell car won't be on the market for at least fifteen years. In the meantime, some 300 million new gasoline vehicles will hit America's highways. They will have a lifetime of at least ten years. It could be twenty-five or thirty years before fuel-cell technology comes to our rescue and radically reduces our dependence on foreign oil. That's too long to wait to start addressing the global oil problem.

How Much Carbon Is Too Much?

When scientists consider the history of world energy use, they talk about decarbonization. Simply put, this is the progressive change in the ratio of carbon to hydrogen atoms used by each successive energy source. For thousands of years, wood was humanity's main source of fuel. It has a ratio of ten carbon atoms to each hydrogen atom. When coal replaced wood as the primary fuel, the ratio of carbonization dropped to about one or two carbon atoms to one hydrogen atom. Oil has one carbon atom for every two hydrogen atoms and natural gas has a ratio of one carbon atom to four hydrogen atoms.

According to the path of decarbonization, each successive fuel source the world has embraced has proven more efficient than the last. And each has also released less carbon dioxide into the atmosphere. But humanity has steadily increased the amount of energy it uses—world energy use has grown seventy-fold since the dawn of the fossil fuel era—and this increase has negated the positive effects of decarbonization. Hydrogen contains no carbon atoms, so conceivably, it would complete the path to decarbonization. But the degree to which hydrogen will save society from the ruin of global warming depends very much on how we obtain it.

Hydrogen may be the most plentiful resource in the world

but it is rarely found in pure form. That means hydrogen must be extracted from other substances that contain it, like water, or fossil fuels like coal, natural gas, and oil.

The Bush administration says hydrogen can break America's dependence on Middle East oil while promoting our own energy industries. The trick—dovetailing neatly with the Cheney National Energy Plan to revitalize U.S. energy industries—is to have America's coal, natural gas, and nuclear-power companies produce hydrogen.

Nuclear power creates no carbon emissions and can be easily used to produce hydrogen. But while the administration would like to see a resurgence of nuclear power in America, most of the country's view of this industry is shaped by the 1979 disaster at Three Mile Island and the job performance of Homer Simpson. With the threat of new terrorist attacks on America's energy infrastructure, building new nuclear plants might just prove to be too radioactive for even this administration.

Then there is coal. You can produce hydrogen from coal by heating the coal, which, of course, produces carbon dioxide. And the U.S. has the largest coal reserves in the world. Because of this abundance, coal has been a major energy priority for the government. The Energy Department plans to spend $1 billion over ten years on a project to extract hydrogen from coal. But producing more global-warming gases just to produce non-global-warming hydrogen is somewhat self-defeating. Industry analysts and the administration tout new scientific methods that aim to trap the CO_2 and store it deep underground. But no one really knows how effective this "carbon sequestration" method is, or whether the CO_2 might leach into drinking water or back up to the surface. Whatever its merits, carbon sequestration will be expensive.

Nearly half the hydrogen currently produced in the world is derived from natural gas. This is the cheapest way of making hydrogen but, as natural gas is a fossil fuel, it also releases CO_2 into the atmosphere (though in far smaller amounts than re-forming hydrogen via coal or gasoline).

Natural gas just happens to be the government's main rationale for more domestic oil exploration. Already, 14 percent of U.S. natural gas consumption goes toward generating electricity, and there are over 270 gas-fired power plants set to come on line in the next decade. The U.S. energy department projects that natural gas consumption will grow by over 70 percent by 2020. But given the strong possibility that natural gas production could peak shortly after oil, and given that the largest deposits of natural gas lie in the Middle East, does it make sense to tie America's already fragile electricity power grid to imported natural gas?

There is another way of producing hydrogen that doesn't require any carbon-based sources. Renewable fuel technologies like solar, wind, hydro, geothermal, and biomass produce hydrogen through electrolysis—the process that splits water into hydrogen and oxygen. As no carbon is released into the atmosphere, the process creates no harmful emissions.

Despite NIMBY resistance in places like Cape Cod and West Virginia, where wind generators have been criticized as eyesores, wind farms will play an increasing role in meeting America's and Europe's electricity needs in the near future, especially as they become more efficient and cost-effective against fossil fuels. Wind power currently accounts for 1 percent of America's electricity needs but the U.S. wind-energy industry expects its business to expand five-fold by 2020. In Europe, wind power is being embraced by nations like Ireland and the UK, which aims to have wind turbines supply 10 percent of the country's energy needs as early as 2010. To realize this goal, the British government has issued wind-generating licenses to companies that would increase the wind-power industry tenfold. The plan calls for a swath of wind turbines up and down the British coast, producing up to 5,000 megawatts of power.

Water-driven renewable energy is beginning to make a splash both in the U.S. and around the world. Countries like Iceland are already proving the effectiveness of hydro and geothermal harnessing. In the U.S., alternative energy experts say the hydropower potential of the Columbia River—enough hydrogen flows through the water every second to fuel 600,000 cars for twenty-four hours—could turn the Pacific Northwest into "the Saudi Arabia of hydrogen."

As always, the stumbling block for renewable energy is its cost. But prices would plummet if the government decided to help renewable energy in the same way they do traditional fossil fuels. The wind-power industry already receives a tax break from the U.S. government—without it, the industry would collapse—but it is a pittance compared to the $14 billion in tax breaks that were proposed for the oil and gas industry in the 2003 Energy Bill.

Blueprint for a New Energy Future

The oil industry has dominated global affairs for nearly 150 years. Most of that time, it has been controlled by a small cadre of powerful companies and producing nations that have set prices and production levels to maximize their profits. How the industry runs its business remains a mystery to most Americans. Gas prices rise and fall, heating-fuel costs spike and the whole northeast of the country can lose power for days on end—yet U.S. consumers remain powerless to do anything about it. Whatever happens to gas prices, we still have to fill our cars and heat our homes. As for the electrical grid: well, it's not like any of us can start a consumer boycott of electricity or start producing our own power.

But with hydrogen drawn from renewable energy resources, we could. It is this potential democratization of our energy supply that is a hydrogen economy's greatest promise. Imagine an energy future where we not only break our dependence on oil, we also have an energy network that is no longer controlled by a small cartel of corporations.

Hydrogen's supporters have a vision of individual energy self-sufficiency. The vehicle they believe could help make it a reality would be the fuel-cell car. A fuel cell is just a big battery, and when its energy is not being called on to power the automobile, it can be used for other things. A fuel-cell vehicle will have a generating capacity of 20 kilowatts. Most cars sit parked over 90 percent of the time. During that idle time they could be "plugged in" to the office, the home, or perhaps even an interactive community or power network—supplying electricity back to the main power grid.

In this way, drivers could sell electricity back to the power grid when they were not using their vehicle. All of the hydrogen would be produced from renewable power sources so there would be no global-warming emissions. And there would be no need to depend on foreign nations for energy.

If so-called distributed generation became a reality, the great blackout of 2003 could not be repeated, and the nation would no longer be dependent on a series of major energy hubs. Yes, the energy grid would be more interconnected than ever—Jeremy Rifkin, author of *The Hydrogen Economy,* calls it a hydrogen energy web—but with the right software infrastructure, the sheer number of energy micro-providers would ensure a twenty-first-century version of Winston Churchill's prerequisite for energy security—"Safety and certainty in oil lie in variety and variety alone."

A diversified hydrogen fuel-cell grid would also have a dramatic effect on the developing world, especially the third of global society that still lives without electricity. Much like the runaway success of wireless telecommunications in Asia, Africa, and South America where land-line telephone systems often don't exist, hydrogen power could change the lives of millions by providing energy that doesn't plunder natural resources or encourage autocratic systems of government.

"By redistributing power broadly to everyone, it is possible to establish the conditions for a truly equitable sharing of the Earth's bounty," writes Rifkin. "This is the essence of the politics of re-globalization from the bottom up."

No one really knows if hydrogen is the answer. But so far, it seems the best option the world has. Industry and government have been slow to embrace hydrogen, but in the next decade, that will likely change, especially as the major nations of Europe and the United States compete to produce an economically viable and independent source of hydrogen fuel. Who succeeds first could also have dramatic repercussions for the future of our environment: will we have a global hydrogen economy that is produced with fossil fuels or carbon-free renewable energy?

The petroleum age is over and we are entering a new era— one where we will harness a new, clean fuel that can meet our energy needs and yet produces no pollution, only water. Hydrogen, if utilized and produced in the right way, offers a great opportunity to break our addiction to oil, strengthen our security, and reverse the environmental damage we are doing to our world. Now is the time to act. It would be a tragedy to see a century of progress be destroyed by a Hundred Year Oil War.

A Note on Sourcing

Throughout this book I have relied on the monthly statistics published by the Energy Information Administration (EIA), an arm of the U.S. Department of Energy, for accurate production and consumption data around the world. BP's annual statistical review and the Paris-based International Energy Agency (www.iea.org), the watchdog for the world's biggest oil-importing nations, were also helpful.

In the preface, my research into all the products that got their start in life as oil came courtesy of the American Petroleum Institute (www.api.org). I traveled to Pennsylvania's oil regions to see Drake's first oil well in Titusville and Oil City. Information about the growth of the oil regions and Pithole, in particular, came from two important books, Daniel Yergin's *The Prize* and Anthony Sampson's *The Seven Sisters*. The story of Oil City's decline came from the *Erie Times-News* (3/27/02). U.S. consumption and production figures came from the EIA.

1. Pursuit of Power: A Short History of Oil

Two sources were invaluable in putting together this chapter, which is essentially a Dummies guide to the history of oil. All the information came either from Yergin's *The Prize,* long considered the bible of oil history, or from an earlier but no less important book, Sampson's *The Seven Sisters.* I owe both authors a debt of gratitude and I hope that, in condensing such a vast subject, I have remained faithful to their interpretation of nineteenth- and twentieth-century history. The *New Yorker* (7/14/03) helped me to understand Iraq's oil history. And *Charlie Wilson's War,* by George Crile, was helpful in explaining the United States' and Saudi Arabia's mutual interests in the Middle East.

2. Car Culture: America's Automobile Addiction

The Prize was also a great help in recounting the growth of car culture in the United States, as was *Asphalt Nation* by Jane Holtz Kay. *The American Gas Station* by Michael Karl Witzel offers a great chronicle of gas stations in twentieth-century American society, while Eric Schlosser's *Fast Food Nation* demonstrated the growth of California's car culture. Keith Bradsher's excellent *High and Mighty* was invaluable in helping me understand America's fascination with the sports utility vehicle (SUV). Most recent facts and figures on U.S. automobile ownership came from a *New York Times* story (8/30/03). The Sierra Club (www.sierraclub.org), the Natural Resources Defense Council (www.nrdc.org), and the Union of Concerned Scientists (www.ucsusa.org) provided a good primer on SUV mileage and safety concerns. Information on the popularity of GM's H2 Hummer came from a *New York Times* story (4/6/03).

3. Exploration or Exploitation?: Oil, Human Rights, and the Environment

In January 1997 I traveled into the Ecuadorian rain forest to research oil pollution. The interviews with Guillermo Maldonado and Vincente Alban took place during that trip and resulted in a story for the *Village Voice* ("Fools' Gold," February 1997). I returned to the Oriente in 1999 and visited indigenous communities in the south of Ecuador that were fighting against oil exploration. Both of these trips provided me with the background to write this chapter. In reconstructing and updating the history of oil exploration in Ecuador, I have been lucky to call upon two expert organizations. Chris Jochnick, founder of the Center for Social and Economic Rights (www.cesr.org), helped me understand the social and economic impact of oil production in Ecuador. And the folks at Amazon Watch (www.amazonwatch.org) provided me with a great deal of up-to-date information on the trial against Texaco and helped me track the history of Ecuador's indigenous resistance. Joe Kane's *Savages* remains the most powerful piece of writing I have read on oil in the Amazon, and it was informative in understanding how oil companies have dealt with indigenous communities in the Amazon. A special report by the *Sacramento Bee* (4/27/03) also captured the dynamics of oil politics in Ecuador. Information regarding Texaco's drilling practices in the region came from research provided by the lawyers representing the plaintiffs in the case.

Information about Dutch disease and the social and economic impacts of oil exploration on communities around the world came from a number of sources, including the work of Professor Michael Ross who was kind enough to give me his time. His paper "Does Oil Hinder Democracy?" can be found in the World Bank's online library (http://econ.worldbank.org/programs/conflict/library/doc?id=21728). Paul Collier's research on war and natural resources for the World Bank was also particularly helpful (some of his work can be found at http://econ.worldbank.org/prr/CivilWarPRR/).

Much has been written about Shell in Nigeria and the death of Ken Saro-Wiwa. Andy Rowell wrote some of the first stories on Saro-Wiwa's trial and death for the *Village Voice* in 1995, and I was lucky to edit those stories. Reports by Project Underground, the Sustainable Energy and Economy Network (www.seen.org), as well as countless newspaper articles informed my retelling of oil's ugly legacy in Nigeria. Information on Nigeria's oil production and revenues comes from the EIA.

Rainforest Action Network has provided me with a great deal of information over the years. In 1997 I traveled with RAN members into the Peruvian Amazon to study oil exploration in the Camisea region and RAN also provided me with information about the U'wa's fight against Occidental in

Colombia. Research on the Publish What You Pay initiative (www.publishwhatyoupay.org) came from Catholic Relief Services, whose report on oil in Africa, "Bottom of the Barrel," is available on its web site (www.catholicrelief.org) and Global Witness (www.globalwitness.org).

4. Boom and Bust: The Price of Oil and How Much Remains

The New York Mercantile Exchange (NYMEX) press relations office was helpful in explaining the convoluted world of oil futures. The EIA (www.eia.doe.gov) also has good research on oil prices and the various contracts that are traded. Peter Beutel at Cameron Hanover was very gracious and informative in fielding my questions about oil prices and how they affect the U.S. economy. And when it came to understanding how crude oil prices translate into the price we pay for gas at the pump, the American Petroleum Institute's October 2002 review of gasoline prices had clear explanations. The *Wall Street Journal* (3/17/03 and 12/04/03) was an invaluable resource on OPEC and oil prices, as was a special report on OPEC in the *Economist* (10/25/03). The *New York Times* (3/2/03) helped explain the correlation between high oil prices and economic recession. Coverage by *Business Week,* the *Guardian,* and the *New York Times* outlined the growth of Russian oil and Western companies' interest in it. Jeremy Rikfin's book *The Hydrogen Economy* was a great introduction to the concept of peak oil production. Further reading of

Colin Campbell's ideas, an interview with Jim Meyer at the Oil Depletion Analysis Center in London, the Association for the Study of Peak Oil and Gas's "2002 Statistical Review of World Oil and Gas," and Kenneth S. Deffeyes's *Sliding Down Hubbert's Peak* helped flesh out the arguments of those who believe global oil production is about to peak.

5. Energy Wars: Oil and National Security

Every time that I thought I was finished with this chapter, something new happened. Such is the peril of writing about geopolitics and current events in book form. Daily coverage of Iraq by the *New York Times,* the *Wall Street Journal,* and the *Guardian* along with weekly issues of *Business Week, Time,* the *Economist,* and the *New Yorker* proved crucial in helping me keep up with the shifting sands of the U.S. occupation of Iraq. George Crile's *Charlie Wilson's War* offered a good analysis of the Carter Doctrine as, of course, did Yergin's *The Prize.* Anthony Sampson's *The Seven Sisters* helped throw light on the 1953 CIA coup in Iran.

Fortune (9/16/02 and 10/14/02) offered smart and detached analysis of how the price of oil, rather than oil itself, affects national security. The *Guardian* (3/11/03) introduced me to the ideas of the Project for a New American Century and its influence on members of the Bush administration. Quotes from the Bush-Cheney National Energy Policy came from the plan itself (www.whitehouse.gov/energy/). Vice

President Dick Cheney's 2002 speech in Tennessee was reported in the Associated Press (8/26/02).

Donald Rumsfeld's 1980s trip to Iraq on behalf of the Reagan administration was first reported by Steve Kretzmann and Jim Valette in "Crude Vision," a March 2003 research paper by the Sustainable Energy and Economy Network, part of the Institute for Policy Studies (www.seen.org). The report also shed light on the UN Oil-for-Food program and its replacement, the Development Fund for Iraq. Paul Collier's World Bank report and Michael Ross's writing on oil and democracy helped my evaluations of oil's ability to rebuild a modern Iraqi society. Information about Iraq's oil potential comes from the EIA Iraq profile, while *Time* offered a good primer on oil politics in the Persian Gulf (5/10/03). The report on Iraq's oil industry by the Council on Foreign Relations and the James A. Baker III Institute for Public Policy of Rice University can be found on the Rice University Web site (www.rice.edu/projects/baker/index.html). Statistics on Saudi Arabia's economy came from the EIA country profile.

Michael Klare's *Resource Wars,* as well as "Global Petro-Politics: The Foreign Policy Implications of the Bush Administration's Energy Plan," published in the March 2002 issue of *Current History,* offered a cogent overview of the role oil will play in shaping U.S. foreign policy. The *Wall Street Journal* (5/27/03 and 6/10/03) offered specifics on the redeployment of U.S. troops. A story in the *Christian Science Monitor* (2/5/03) threw light on how U.S. forces were training the Colombian army to protect oil pipelines as did a *New York Times* story (10/4/02). Information on members of the U.S.–Azerbaijan Chamber of Commerce came from its Web site (www.usacc.org).

Wired writer Richard Martin provided good background information on current Caspian oil projects. Transcripts from an April 2003 meeting of the Senate Foreign Relations Committee subcommittee on international economic policy were also helpful in regard to Caspian oil. Dick Cheney's belief in encouraging ties with Iran was written about in the *Houston Chronicle* (11/28/97). Hamid Karzai's role as negotiator for Unocal was reported by the *Guardian* (10/11/03). Information on Caspian pipelines and the Caspian Pipeline Consortium came from many sources but the *New York Times* (11/10/97), a special project by Friends of the Earth (www.foe.org), and *Business Week* (12/24/01) were particularly helpful.

The role of U.S. forces in Georgia was discussed in the London *Observer* (5/12/2002) and China's Caspian deals were the focus of a *Business Week* story (3/31/03). The *Nation* wrote extensively about Equatorial Guinea's oil industry (4/22/02) and the *New Yorker* ran an excellent piece about São Tomé in October 2002. The *Los Angeles Times* offered good insight on oil politics in Africa (1/13/03), as did *Time* magazine (12/15/02), *Le Monde Diplomatique* (1/8/03) and the *Economist* (1/23/03). Information on the African Oil Policy Group and the push in Congress to increase African oil imports came from the State Department Web site and from coverage in the *Guardian* (7/7/03). Nigeria and São Tomé's commitment to greater transparency in the oil contracts they sign was noted by Global Witness.

6. Oil in the Family: The Bush Administration's Energy Policy

George W. Bush's family history has been told in many places. Specifics on his oil industry experience came from *Salon* (11/19/01). Information on the Bush administration's oil industry experience came from the NRDC (http://www.nrdc.org/air/energy/aplayers.asp). A Common Cause report ("Drilling for Bargains," www.commoncause.org) and the Center for Responsive Politics (www.opensecrets.org) provided a lifeline in detailing the history of political contributions from the oil and automobile industries. All information on the National Energy Policy came from the plan itself (www.whitehouse.gov/energy). The energy plan also outlined the administration's case for new oil and gas exploration in the Arctic National Wildlife Refuge and federal lands in the lower forty-eight states. The Sierra Club ("Crude Behavior Report," www.sierraclub.org) and research by Cutler J. Cleveland and Robert K. Kaufmann, of the Center for Energy and Environmental Studies and the Department of Geography at Boston University (http://www.hubbertpeak.com/Cleveland/bushpolicy.htm) offered arguments against new exploration in Alaska. The *Anchorage Daily News* published a report on the environmental impacts of oil exploration in

Alaska (3/5/03). And *The Hydrogen Economy* also offered interesting figures on how much oil could be considered recoverable in ANWR.

The *New York Times*, (2/8/02, 11/22/02, 12/29/02), the *Los Angeles Times*, (2/2/03), *USA Today*, (3/19/02), and *Travel and Leisure* (04/03) all recounted the Bush administration's push for new oil and gas exploration in Texas and the western Rockies. The *New York Times* also wrote extensively about Interior Secretary Gale Norton's plans for America's wilderness areas, and reports by The Wilderness Society (www.wilderness.org) offered a detailed analysis of the oil and gas industry's access to public lands.

7. CAFE Society: The Brief Life and Untimely Death of Energy Conservation

Telling the history of U.S. fuel-economy legislation would have been a thankless task were it not for Keith Bradsher's book, *High and Mighty,* which made auto industry lobbying and Capitol Hill legislative infighting seem interesting. In February 2002 an alliance of non-governmental groups, including the NRDC, the Sierra Club, and the Union of Concerned Scientists, published a report titled *Increasing America's Fuel Economy*. It proved useful in understanding the complexities of CAFE standards. The Almanac of Policy Issues also offered a good primer on fuel-economy standards. The 2001 National Academy of Sciences report on fuel-economy standards can be found at www4.nationalacademies.org. A Heritage Foundation report (7/11/01) argued that CAFE standards threatened U.S. automobile standards and made drivers less safe. And a report by the American Institute of Physics argued that greater fuel-economy standards did not result in automobile safety risks.

Time magazine (2/23/03) offered SUV culture as its cover story. "The Detroit Project" can be found at www.ariannaonline.com/suv/, and the Sierra Club's "Hummer Hit" can be found on its Web site. The Rev. Jim Ball's SUV campaign was reported by the *Detroit News* (6/29/03). Rainforest Action Network has details of its Ford Motors campaign on its Web site (www.ran.org). The *New York Times* wrote about the SUV torching campaign in California (8/31/03). The polls showing support for tougher fuel-economy standards were conducted by Lake, Snell, Perry & Associates for the Sierra Club (www.sierraclub.org/scoop/poll.asp). The 2003 Bush administration initiative to reconsider fuel-economy standards was reported by the *Wall Street Journal* (12/22/03 and 12/23/03). Daniel Becker's quote comes from a Sierra Club press release in response to the plans.

The Union of Concerned Scientists offered an in-depth report on hybrid vehicles (www.ucsusa.org), and the *Christian Science Monitor* (2/24/03) considered the price discrepancy between gasoline and hybrid vehicles. The JD Power and Associates study can be found at www.jdpower.com/awards/industry/summary.asp?StudyID=611. A *Business Week* special report ("Getting Smart About Oil," 2/24/03) augmented this research on hybrids and the *Wall Street Journal* discussed the auto industry's embrace of diesel engines (7/28/03). NRDC's 2003 "Dangerous Addiction" report offered a five-step plan to breaking oil dependence using better fuel-economy standards and hybrids. The American Petroleum Institute and the EIA both provided detailed breakdowns on the retail price of gasoline. As for the real price of gasoline, the International Center for Technology Assessment report on the true social and economic costs of gas can be found at http://www.icta. org/projects/trans/rlprexsm.htm. *Mother Jones* also tackled this subject (10/5/01) and the *Christian Science Monitor* (3/13/03) considered how high gas prices affect California motorists. *Business Week*'s special report (2/24/03) addressed the thorny issue of raising gasoline taxes.

8. Beyond Oil: The Hydrogen Balm

In the summer of 2003 I traveled to Dearborn, Michigan to attend Ford's centennial celebrations. I missed Beyoncé, but I did get to tour the new Rouge factory and see the green roof. As for the oil companies' green strategies, information about BP's renewable projects came from the company Web site. Lord Browne's comments on BP's marketing campaign were part of a power lunch titled "Beyond Fossil Fuels" organized by *Sierra* magazine. Information about BP's Russian ventures was published in the *Wall Street Journal* (11/13/03) and the London *Observer* (10/19/03). Canada's tar sand potential was discussed in a *New York Times* Op-Ed (8/14/03) and in an article titled "Canada Builds a Large Oil Estimate on Sand" (6/18/03).

The FreedomCAR initiative is explained at www.eere. energy.gov/hydrogenfuel. The history of hydrogen was recounted in the *New Yorker* ("The Car of Tomorrow," 8/11/03) as was information about GM's HydroGen project. How a fuel cell works and the potential difficulties in making it commercially viable are discussed in an April 2003 *Wired* article ("How Hydrogen Can Save America") and in the *Wall Street Journal* (03/07/03). The *New Yorker* (8/11/03) also provided information on Iceland's hydrogen strategy. NRDC's 2003 "Dangerous Addiction" report critiqued the Bush administration's hydrogen strategy. *Wired,* the Union of Concerned Scientists, and *The Hydrogen Economy* all offered astute explanations of how hydrogen is produced. The Department of Energy's Web site (www.energy.gov) explained carbon sequestration and the *Seattle Post-Intelligencer* (7/8/03) wrote about renewable fuels in the Pacific Northwest.

The *New York Times* (8/28/03) outlined wind power's potential as a part of America's energy production. The *Guardian* (7/14/03) discussed the British government's plans to increase wind power production in the UK. Rifkin's *The Hydrogen Economy* and a cover story of the same name in *E* magazine helped sketch out the notion of a hydrogen energy web and the possibility of an energy future independent of fossil fuels.

Afterword: U.S. and China Oil

A lot has happened in the world of oil since I first sat down to write this book in 2003. President George W. Bush won re-election for one thing. U.S. troops remain in Iraq over two years after the war began. And as for those weapons of mass destruction—well the more or less said about them the better, depending on your political point of view.

Still, there is one oil undercurrent that resonates more than any other. Since 2003 the price of oil has been rising, soaring, even; first through the $40 mark then blasting its way through $50 a barrel in October 2004 before flirting with $60 a barrel. As I write now, the price of crude for sales on the New York Mercantile Exchange (NYMEX) sits at $58 a barrel. The days of cheap oil seem a long way away.

Just two years ago, the idea that oil-dependent western societies would have to tolerate oil prices at $50 a barrel scared the hell out of economists and economic ministers alike. Fifty dollars a barrel was a mythical ceiling—if oil prices stayed above that level for any prolonged period of time, it was asserted, the global economy would falter and the inevitable impact of lessening global demand would bring prices back down to a more sensible and natural level.

That didn't happen; crude prices first hit $50 in July of 2004 and, give or take a few weeks when they've dropped back into the $40 range, they've remained in this supposed economic red zone for well over a year.

At first, oil observers attributed the price spike to the so-called risk premium that traders were placing on oil supplies as a result of the war in Iraq and the knock-on instability in the Middle East. There was also the seedy role played by speculators, who, it was said, were driving up the price of oil by taking advantage of a volatile short-term market.

Both explanations were correct, up to a point. The biggest factor in the price rise, however, had more fundamental roots. The price of oil is now firmly entrenched at $50-plus a barrel because demand for the black stuff continues to outstrip the supply of crude available. What's more—and this is one prediction I'm willing to go out on a limb on—the price of oil is going to stay around that level for years to come.

There are many reasons to take this view. One is that the oil industry doesn't have enough refinery capacity to meet the growing demand for gasoline. A second more contentious and far less provable piece of thinking is that most of the world's easily extracted oil has already been plundered and that the current price rise represents the beginning of a final period of oil production that will cost more and more to bring to market, and hence cost more for you and me to put into our cars.

By far the most persuasive argument however is that, despite oil prices hitting $50 a barrel, the new price threshold has done little to dampen global demand. Quite the contrary. While it's true that industrialized nations like the United States have become more energy efficient and hence less reliant as a whole on oil to power their economy, there is an exception. Most oil in the U.S. is used for transportation. While makers of SUVs are taking a bit of a hit, for the most part U.S. consumers have soaked up the higher price of gasoline into their weekly expenditures and kept on driving. Indeed, U.S. oil consumption is up 3 percent over last year, and 2004 was the thirstiest year yet recorded.

But while the U.S. and Europe continue to crave more oil, parts of the developing world hunger for it even more. There, feverish industrialization looks likely to double the global demand by 2030.

Leading the charge is China. In 2003, the Chinese economy grew 9.1 percent and over two million Chinese bought their first automobile. In 2004, that growth continued and China overtook Japan as the world's second largest oil consumer, sending global oil prices skyrocketing as a result of its increased demand. By mid 2005, China's demand for oil had increased an incredible 65 per cent in just three years. By 2020, it is expected to replace the United States as the world's leading consumer of oil.

At the start of this book, I identified the triple threat to the American geo–body politic posed by our addiction to oil. But China's breakneck economic development, accompanied by its forthright pursuit of oil supplies all over the world, is certain to compound the problems faced by the United States and all other major oil dependent nations. With a population four times the size of the United States, and with an economic growth strategy and geopolitical outlook that makes nineteenth-century American Manifest Destiny look humble in comparison, China's rush to satisfy its energy craving over the next decade is going to play a central role in dictating the well-being of our global economy, the security of our nations, and the health of our planet.

Unlike the quaint old industrialization of the eighteenth, nineteenth, and twentieth centuries—which saw nations develop a heavy industry base, then a manufacturing base, and finally a high-technology crème de la crème—China seems intent on doing all of this at the same time. Such is the power of its employee-based foundation that China is able to excel in heavy industry while also whipping the world in manufacturing flexibility. It is even threatening to give the U.S. a run for its money in the knowledge-industries with its intellectual and technological expertise.

The upshot of this industrial charge is two-fold.

First, as more and more heavy industrial plants, manufacturing centers, and high-tech R&D clinics pop up all over China, so the authorities must build a power infrastructure to provide them with electricity. At present a great deal of China's oil needs come from power generation. Just like the United States, China is a major oil producer in its own right: in 2004 it was ranked sixth in the list of global oil producers, just three places below the U.S. And just like the U.S. in the past, it based its power generation strategy on its oil strength. Over thirty years have passed since the United States last built an oil-fueled electricity power plant—around the same time that it realized the U.S. domestic oil industry couldn't match America's own oil needs. China, you guessed it, also can't match its economy's demand for oil and so has been importing steadily more crude to keep its industrial growth on track.

China's future economic growth might see the nation become more efficient in using its oil resources to produce electricity just as the U.S. did. But harnessing other energy sources to create electricity poses its own problems.

Like the United States, China is considering natural gas, nuclear, and coal for its electricity needs. China has vast quantities of coal and it seems keen to build its own new generation of coal power plants. But while the U.S. plans to pioneer new clean-coal technology that captures CO_2 emissions before they enter the atmosphere and then stores them securely deep underground, the new generation of Chinese power plants are not likely to be fueled by clean coal. The resulting greenhouse gases generated by China's drastic need for new electricity should be a major worry for the entire world.

Power generation still requires a lot of oil but, as all Western nations know, not nearly as much as a transportation system dominated by automobiles. Just five years ago, car trouble was not something China worried much about. Today, a burgeoning middle-class, flush with the salaries that accompany a national economic gallop, have abandoned the bicycle—long the symbol of a classless workforce—for the four-wheeled one-upmanship of a motor car.

The result is both amusing and also deeply disturbing. Downtown Beijing, along with most other Chinese cities nowadays, is a driving instructor's idea of hell. Thousands upon thousands of evidently licensed yet plainly incompetent novice drivers have taken to the streets in their Chinese-manufactured, black VW Passat clones figuring that practice and more practice is the best way to learn how to drive.

More experienced motorists—often official drivers for the rich or major corporations—carve a high-speed path through the auto-didacts in their Mercs and Beemers en route to the pristine new highways the Chinese government is building to encourage this transport revolution.

Drivers need cars and U.S. automakers, led by Ford and GM, have their pedals to the metal in a rush to capture just part of the nascent China car market. China is already thinking fast and hard about taking advantage of hybrid energy technology to mitigate its citizens growing dependence on foreign oil. But even as it does so, China has also embarked on a global hunt to secure new sources of oil—one that is likely to result in time in a head-on clash with the only other power that consumes more oil than itself: the United States.

Ever since the First World War, major nations have sought oil security. Acutely aware of its oil vulnerability, the Chinese premier Hu Jintao has dispatched the state-run oil companies on a spending spree to snap up oil supplies all around the world. From Algeria to Venezuela, China is cutting oil production and import deals with major and minor oil-producing countries in Africa, the Persian Gulf, Central Asia, and even America's backyard, Latin America and Canada.

It's the same oil diversification strategy that the U.S. employed in the early 1980s after the OPEC price shocks exposed its own dependence on foreign oil. Now, the two largest oil importers are chasing the same sources of the oil in the name of energy security.

In all the major oil-producing regions of the world, China and the U.S. are engaged in a potentially dangerous game of building oil alliances.

Few oil exporters lack attention from at least one of the two suitors, both of whom favor promiscuity in order to insure against local disruptions in oil supply. But there is a certain menace to this ardor as well, as reflected by the unerring flow of troops and arms to oil-rich regions.

Although the U.S. strategy is based not on physically controlling oil fields but, rather, on maintaining a stable price for crude oil on the global market, oil-producing regions are thick with U.S. military bases—and in Nigeria, Kazakhstan, and several other countries, U.S. troops are working with local militaries to protect oil infrastructure. Military aid is also used to woo tinpot oil regimes, opening the door for U.S. companies such as ExxonMobil and ChevronTexaco to invest in local fields.

Late to the game and largely unable to compete with the U.S. military, China has sought to meet its energy needs by making use of its newfound economic clout—most notably by buying drilling and refining rights throughout the world. Its state-owned oil companies have spent hundreds of billions of dollars on such rights, often at high premiums; since 1993 China has invested in more than fifty oil and gas projects in some thirty nations. It has sought to exploit power vacuums in the Caspian region where Russian influence is on the wane and has been successful in securing a foothold in Kazakhstan's oil industry—expected to be the world's fifth largest by 2010. And China has focused on acquisitions and partnerships in pariah nations—Sudan, Iran—where the United States has refused to tread.

In Sudan alone China has reportedly spent $15 billion developing oil fields. Yet China, too, has begun to use its military to protect its investments and—indirectly—to open new oil markets. Reportedly, troops disguised as oil workers patrol Chinese oil infrastructure in Sudan. And in recent years China has consolidated its military presence in oil- and gas-rich parts of the South China Sea, sovereignty over which is disputed.

In 2004, a Chinese nuclear submarine entered Japanese territorial waters in what was widely seen as saber rattling over ownership of natural-gas fields and potential oil reserves in the East China Sea. China and other countries have also erected territorial markers around the Spratly Islands, in the South China Sea, which are rumored to hold billions of barrels' worth of oil and gas. China's offshore oil claims, and its concern that a conflict with Taiwan would push the U.S. Navy to close access to the Strait of Malacca (through which

60 percent of Chinese oil imports flow), may partly explain China's recent naval expansion as well.

China has also been busy in America's own backyard. In return for cooperation on oil projects, China has lent support to Brazil's bid for a UN Security Council seat. Meanwhile, President Hugo Chavez, of Venezuela, eager to reduce its dependence on the United States market, has met with Chinese government officials and invited Chinese oil companies to explore Venezuela's oil fields and build refineries. He has also proposed new pipelines to the Pacific that would make oil shipments to Asia cheaper. China has even set about wooing Canada for a piece of the vast Alberta tar sand reserves, second only to Saudi Arabia in recoverable oil but with production costs more than ten times as high.

In June 2005, China announced its energy strategy in terms corporate America could really relate to: the state-run China National Offshore Oil Company (CNOOC) launched an un-solicited bid to buy the California-based oil mini-giant, Unocal. Most of Unocal's oil assets are based in Asia, far closer to the China than the U.S., so the deal made sense in pure geographic terms. But China's move to control a U.S. oil company only underlined its global ambitions and its hunger for more and more oil. The bid sent shockwaves through both Wall Street and Washington.

Perhaps most significant in the short term is China's relationship with Iran. It is a matter of simple mathematics that neither China nor the United States will be able to avoid reliance on the Middle East, where three countries—Saudi

Arabia, Iraq, and Iran—sit on nearly half of the world's easily accessible oil. With Saudi Arabia and Iraq clearly within the U.S. sphere of influence, China has been steadily courting Tehran and aims to become the biggest buyer of Iranian oil. In return for oil, China has supplied weapons—most notably anti-ship missiles, which Iran has aimed at the Strait of Hormuz, the busiest oil-export route in the world—and technology and materials that can be used for the manufacture of nuclear weapons.

Will oil concerns inside or outside the Middle East lead to open conflict between the United States and China? In the short run that's unlikely; even if the United States were to invade Iran, China probably couldn't do anything about it. But given oil's historical place in military conflicts between great powers, one cannot discount the possibility that over time, as demand for oil increases and supplies of crude become harder and more expensive to tap, the two nations will come to blows.

In the meantime, China will probably continue to offer succor to rogue states that America would like to see isolated. And weapons of both Chinese and U.S. origin will continue to flow into regions where oil itself has already bred government corruption, vast income inequality, and war.

—Matthew Yeomans
August 2005

Index